# THE REAL PROJECTIVE PLANE

By the same author

NON-EUCLIDEAN GEOMETRY
University of Toronto Press

REGULAR POLYTOPES
Methuen, London

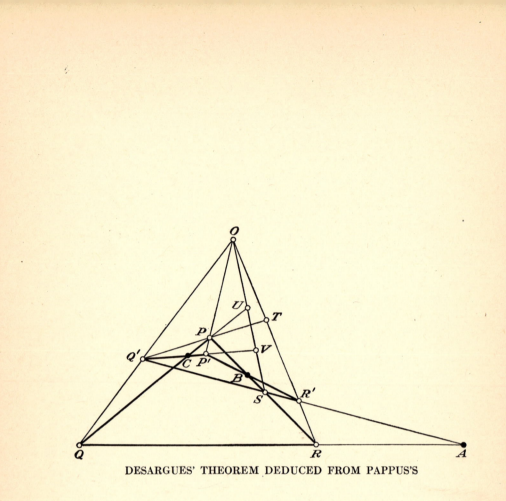

DESARGUES' THEOREM DEDUCED FROM PAPPUS'S

# THE REAL PROJECTIVE PLANE

By

## H. S. M. COXETER

*Professor of Mathematics*
*University of Toronto*

**FIRST EDITION**

NEW YORK    TORONTO    LONDON

## McGRAW-HILL BOOK COMPANY, INC.

1949

THE REAL PROJECTIVE PLANE

THE MAPLE PRESS COMPANY, YORK, PA.

# PREFACE

This introduction to projective geometry can be understood by anyone familiar with high-school geometry and algebra. The restriction to real geometry of two dimensions makes it possible for every theorem to be illustrated by a diagram. The early books of Euclid were concerned with constructions by means of ruler and compass; this is even simpler, being the geometry of the ruler alone. The subject is used, as metrical geometry was by Euclid, to reveal the development of a logical system from primitive concepts and axioms. Accordingly the treatment is mainly synthetic; analytic geometry is confined to the last two of the twelve chapters.

The strict axiomatic treatment is followed far enough to show the reader how it is done, but is then relaxed to avoid becoming tedious. Continuity is introduced in Chapter 3 by means of an unusual but intuitively acceptable axiom. A more thorough treatment is reserved for Chapter 10, at which stage the reader may be expected to have acquired the necessary maturity for appreciating the subtleties involved.

The spirit of the book owes much to the great *Projective Geometry* of Veblen and Young. That dealt with geometries of various kinds in any number of dimensions; but the present book may be found easier because one particular geometry has been extracted for detailed consideration. Chapters 5 and 6 constitute what is perhaps the first systematic account in English of von Staudt's synthetic approach to polarities and conics as amplified by Enriques: A polarity is defined as an involutory point-to-line correspondence preserving incidence, and a conic as the locus of points that lie on their polars, or the envelope of lines that pass through their poles. This definition for a conic gives the whole figure at once and makes it immediately self-dual, a locus and an envelope, whereas Steiner's definition assigns a special role to two points on the conic, obscuring its essential symmetry. Moreover, the restriction to real geometry makes it desirable to consider not only the hyperbolic polarities which determine conics but also the elliptic polarities which do not. The latter are important because of their application to elliptic geometry. (In complex geometry this distinction is unnecessary, for an elliptic polarity determines an imaginary conic.) The linear construction for the polar of a given point (5·64) was adapted from a question in the Cambridge Mathematical Tripos, 1934, Part II, Schedule A.

The treatment of conics is followed in Chapter 8 by a description of affine geometry, where one line of the projective plane is singled out as a *line at infinity*, enabling us to define parallel lines. It is interesting to see how much of the familiar content of metrical geometry depends only on incidence and parallelism and not on perpendicularity. This includes the theory of area; the distinction between the ellipse, parabola, and hyperbola; and the theory of diameters, asymptotes, etc. The further specialization to Euclidean geometry is made in Chapter 9 by singling out an *absolute involution* in the line at infinity.

Chapter 10 introduces a revised axiom of continuity for the projective line, so simple that only eight words are needed for its enunciation. (This has not been published elsewhere save as an abstract in the Bulletin of the American Mathematical Society.) Chapter 11 develops the formal addition and multiplication of points on a conic and the synthetic derivation of coordinates. Finally, Chap. 12 contains a verification that the plane of real homogeneous coordinates has all the properties of our synthetic geometry. This proves that the chosen axioms are as consistent as the axioms of arithmetic.

Almost every section of the book ends with a group of problems involving the latest ideas that have been presented. All the difficult problems are followed by hints for solving them. The teacher can render them more difficult by taking them out of their context or by omitting the hints.

I take this opportunity for expressing my thanks to H. G. Forder and Alan Robson for reading the manuscript and suggesting improvements; also to Leopold Infeld and Alex Rosenberg for helping with the proofs.

H. S. M. Coxeter

Toronto, Ont.
*February,* 1949

# CONTENTS

## CHAPTER 1

# A COMPARISON
# OF VARIOUS KINDS OF GEOMETRY

**1·1  Introduction.**  The ordinary geometry taught in high school, dealing with circles, angles, parallel lines, similar triangles, and so on, is called *Euclidean geometry* because it was first collected into a systematic account by the Greek geometer Euclid, who lived about 300 B.C. His treatise, *The Elements*, is one of the most famous books in the world; probably the Bible is its only rival in the number of copies made and the number of languages into which it has been translated. With a few unimportant changes it is still suitable for the instruction of the young.

During the nineteenth century there was a tendency to extract from Euclidean geometry certain ideas of a particularly simple nature, especially ideas that did not involve measurement of distance or angle, and to use these for building up more general systems, notably *affine* geometry and *projective* geometry. The meaning of these terms will be clear after we have examined certain kinds of projection. For that purpose we shall need some intuitive notions of solid geometry; but after the present chapter we shall be concerned solely with plane geometry.

These new systems are said to be more general because, besides throwing fresh light on Euclidean geometry itself, they are capable of extension in other directions by the introduction of new kinds of measurement. Affine geometry can be developed into Minkowski's geometry of the space-time continuum considered in the special theory of relativity, and projective geometry can be developed into the various kinds of "non-Euclidean" geometry that are relevant to more modern ideas of relativistic cosmology. This remark is intended merely to show why it is worth while to study these fundamental geometries; the extensions themselves are beyond the scope of this book.

**1·2  Parallel Projection.**  Two figures in distinct planes are said to be derived from each other by *parallel projection* if corresponding

1

points can be joined by parallel lines.* (This is essentially what happens when the sun casts a shadow on the ground; *e.g.*, when a circular coin casts an elliptic shadow, the lines joining each point of the circle to its shadow on the ellipse are parallel.) If the two planes are parallel, the two figures will be exactly alike (congruent); otherwise they may have somewhat different shapes, but straight lines remain straight, tangents to curves remain tangent, parallel lines remain parallel, bisected segments remain bisected, and equal areas remain equal. In other words, the properties of straightness, tangency, parallelism, bisection, and equality of area are *invariant* under parallel projection. Such properties are the subject matter of *affine* geometry. (This use of the word *affine* is due to the Swiss mathematician Euler, 1707–1783.)

On the other hand, the content of *projective* geometry is still more restricted, being confined to those properties (such as straightness and tangency) which remain invariant under central projection.

**1·3 Central Projection.†** Two figures in distinct planes are said to be derived from each other by *central projection* if corresponding points can be joined by *concurrent* lines, all passing through a fixed point $L$. (This is essentially what happens when a lamp casts a shadow on a wall or on the floor. The circular rim of a lampshade usually gives a larger circular or elliptic shadow on the floor and a hyperbolic shadow on the nearest wall.) If the two planes are parallel, the two figures will be similar and the invariant geometry will again be affine. So we shall assume the two planes to be nonparallel; then the plane through $L$ parallel to one of the two planes will meet the other in a definite line called the *vanishing line* for a reason that will soon be explained.

Figure 1·3A represents a box standing on a table with a lamp suspended inside the lid at $L$. A figure is drawn opaquely on the transparent vertical side $oP$ of the box so as to cast a shadow on the horizontal plane of the table top. Clearly, the shadow is derived from the original figure by central projection from $L$. In general, two intersecting lines project into two intersecting lines, just as in the case of parallel projection; but an exception arises when the given lines intersect on the special line $o$, which lies in the horizontal plane through $L$. Such lines, say $AP$ and $AQ$, project into *parallel* lines $p$ and $q$ through $P$ and $Q$, both parallel to $LA$. Conversely, any two parallel lines on the table top are each coplanar with the parallel line through $L$, say $LA$; therefore, unless they are parallel to $o$, they must be projected images of two lines through a definite point $A$ on $o$.

* We shall always use the word *line* in the sense of a straight line of unlimited extent.
† Cremona (Ref. 8, p. 3). (All references are listed in the Bibliography on pp. 189–190.)

This process of central projection relates points in the vertical and horizontal planes in such a way that the join of two corresponding points always passes through $L$. Points in the vertical plane below the table top project into points inside the box, and points $X$ above $o$ project into points behind the box (on $XL$ produced), although now the notion of a shadow breaks down. Thus the only points that have no images are those on the line $o$, which is consequently called the *vanishing line*.

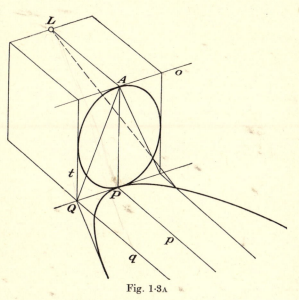

Fig. 1·3A

A circle in the vertical plane is joined to $L$ by a cone (usually an oblique circular cone); thus it projects into a conic section, or *conic*. If the circle does not meet $o$, the conic is a closed oval curve, *viz.*, an *ellipse*. If the circle cuts $o$ in two distinct points, the conic is a *hyperbola*, which has two branches arising from the arcs below and above $o$. (The latter branch is behind the box.) Finally, if the circle touches $o$ (as in Fig. 1·3A), the conic is a *parabola*, which has only one branch but is not closed. Note that other tangents to the circle (such as $t$) project into tangents to the conic. We shall not be surprised to find that the role of conics in projective geometry is almost as vital as the role of circles in Euclidean geometry, though actually we shall not make use of them till Chap. 6.

We have taken the plane $oP$ to be vertical for simplicity. An oblique plane could be used just as well.

These ideas formed the foundation for the work of a remarkable Frenchman, Jean Victor Poncelet (1788–1867), who fought in Napoleon's Russian campaign (1812) until the Russians took him prisoner. Being deprived of all books, he decided to reconstruct the whole science of geometry. The result was his epoch-making *Traité des propriétés projectives des figures*,* which was published in 1822.

### EXERCISES

1. Lines parallel to the vanishing line remain parallel after projection, but any other parallel lines project into intersecting lines. Show how to determine their point of intersection.

2. Let a circle cut the vanishing line in $A$ and $B$. Observe how the tangents at $A$ and $B$ project into the asymptotes of the hyperbola.†

**1·4 The Line at Infinity.** In Sec. 1·2 we defined affine geometry as consisting of those propositions of Euclidean geometry which retain their meaning and validity after parallel projection; thus every proposition of affine geometry holds also in Euclidean geometry, but other propositions of Euclidean geometry (such as Euclid I. 1 and 5) are essentially meaningless in affine geometry. Somewhat similarly, projective geometry includes all propositions of affine geometry that retain their meaning and validity after central projection; but that is not the whole story. Some statements are true in projective geometry but false in affine geometry. The most important instance is: "Any two lines in a plane have a point of intersection." This fails in affine geometry because the two lines might be parallel. The projective statement is validated by inventing a new kind of "point" so as to be able to say that parallel lines have a common *point at infinity*: the projected image of a point on the vanishing line. This vitally important concept is due to the great German astronomer Kepler (1571–1630).

We often think of a line as consisting of all the points on it, *i.e.*, a *range* of points. It is equally useful to think of a point as consisting of all the lines through it, *i.e.*, a *pencil* of lines. Statements about points are easily translated into statements about pencils; *e.g.*, "Two points lie on just one line" becomes "Two pencils contain just one common line." Lines through $A$ (in the plane $oP$ of Fig. 1·3A) project into parallel lines (such as $p$ and $q$) on the horizontal plane. If we agree to call these a *pencil of parallels*, we may say that a pencil always projects into a pencil. When statements about such pencils are translated back into

---

* Poncelet (Ref. 29).

† Readers who have not studied the hyperbola should omit this exercise.

statements about points, we have to admit points at infinity as well as ordinary points. In fact, we call the pencil of parallels a point at infinity, denote it by $A'$, and call it the projected image of the ordinary point $A$ on the vanishing line $o$.

In this manner we have extended the meaning of the word *point* so as to be able to say that any two coplanar lines intersect in a point. Similarly, we extend the meaning of the word *line* so as to be able to say that any two planes intersect in a line. If the two planes happen to be parallel, this is a *line at infinity*. Since we have agreed to call $A'$ the projected image of $A$, all points at infinity in the plane $pq$ lie also in the parallel plane $Lo$ and form a "range" on the line at infinity, which is the intersection of the two parallel planes, *i.e.*, the projected image of the vanishing line $o$. However, there seems to be a paradox here: we have agreed that the point at infinity $A'$ is really only another name for the pencil of lines parallel to $p$, and yet we have declared that it lies in the plane $Lo$, which certainly does not contain these lines. The explanation is that for brevity we have oversimplified the account. For a complete treatment of "ideal elements" we should have to consider the whole space, using *bundles* of lines and planes (*i.e.*, all the lines and planes through a given point or parallel to a given line*) instead of pencils of lines. Then a "point" is said to lie in a plane if the plane belongs to the bundle; and in the case of a bundle of parallels this merely means that the plane contains a line in the direction of the bundle. When we restrict consideration to a single plane, the bundle is replaced by a pencil, all the lines of an ordinary pencil contain different points at infinity (which belong equally to the respectively parallel lines of any other ordinary pencil), and all these points at infinity are to be regarded as a range on the line at infinity. We can treat the points at infinity just like any other range of points so long as we are dealing with properties that are invariant under central projection.

By introducing these new elements we have enlarged the affine plane (in which both the affine and Euclidean geometries operate) so as to obtain the *projective plane*, which has simpler properties of incidence. For other purposes we might choose different ways to enlarge the affine plane, but the corresponding geometries would be outside the scope of this book.

---

* Any two coplanar lines (intersecting or parallel) determine such a bundle, which contains the intersections of all the planes through one line with all the planes through the other [see Coxeter (Ref. 6, p. 166)]. The analytic aspect of ideal elements is neatly described by Hardy (Ref. 15, p. 449).

Which of the following propositions belong to Euclidean geometry, which to affine, and which to projective?

   *a.* Four lines of general position have six points of intersection.

   *b.* If lines *AB* and *CD* intersect, then *AC* and *BD* intersect.

   *c.* The diagonals of a parallelogram bisect each other.

   *d.* The three medians of a triangle have a common point.

   *e.* The three altitudes of a triangle have a common point.

   *f.* The angle in a semicircle is a right angle.

**1·5  Desargues' Two-triangle Theorem.** If the three sides of one triangle are parallel to the three sides of another, the two triangles are,

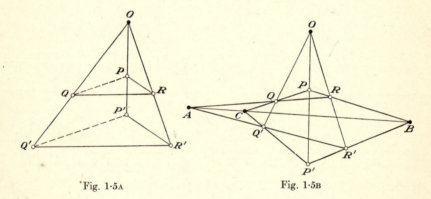

Fig. 1·5A                                    Fig. 1·5B

of course, similar. Thus the part of Euclid's Book VI that deals with similar *and similarly situated* figures belongs to affine geometry. The following theorem is easily proved in this manner:

**1·51**  *Let PQR and P'Q'R' be two triangles (in the affine plane) with QR parallel to Q'R' and RP parallel to R'P', while the joins PP', QQ', RR' are concurrent. Then PQ is parallel to P'Q'.*

*Proof:* Let *O* be the common point of *PP'*, *QQ'*, *RR'*, as in Fig. 1·5A. Applying Euclid VI. 2 to triangles *QOR* and *ROP*, we have

$$\frac{OQ'}{OQ} = \frac{OR'}{OR} = \frac{OP'}{OP}.$$

Hence the triangles *POQ* and *P'OQ'* are similar, and their corresponding sides *PQ* and *P'Q'* are parallel.

This affine theorem has a projective generalization that is very important:

**1·52   Desargues' Two-triangle Theorem.** *If two triangles have corresponding vertices joined by concurrent lines, then the intersections of corresponding sides are collinear.*

In other words, if $PP'$, $QQ'$, $RR'$ all pass through one point $O$, as in Fig. 1·5B, then the intersections

$$A = QR \cdot Q'R', \qquad B = RP \cdot R'P', \qquad C = PQ \cdot P'Q'$$

all lie on one line.

*Proof:* Imagine the figure drawn on the plane $oP$ of Fig. 1·3A, with $AB$ for vanishing line. By projection onto the horizontal plane we obtain two triangles having two pairs of parallel sides, as in 1·51. We conclude that their remaining sides are parallel. Since the sides $PQ$ and $P'Q'$ of the original triangles project into these parallel lines, their point of intersection $C$ must lie on the vanishing line $AB$, as required.

Of course, 1·51 is just a special case of 1·52, obtained by taking $AB$ to be the line at infinity. In projective geometry the line at infinity is treated like any other line; therefore, if the corresponding sides meet in collinear points in this special case, they must still do so for *any* position of $AB$. In this spirit, instead of projecting the figure onto another plane, we could say, "Make the plane affine by choosing $AB$ as the line at infinity."

Since we shall eventually take Desargues' theorem (our 1·52) as an axiom, it seems worth while to give an alternative proof: von Staudt's projective three-dimensional proof. First, we observe that the theorem is almost obvious when applied to two triangles in distinct planes; for in that case the points $A$, $B$, $C$ all lie in the plane $PQR$ and also in the plane $P'Q'R'$, and therefore they all lie on the line of intersection $PQR \cdot P'Q'R'$. The theorem for triangles in one plane arises as a limiting case; but if we prefer not to use such considerations of continuity, we may proceed as follows: Take any two points $S$ and $S'$ on a line through $O$ outside the plane of the two given triangles, so that the four lines $PP'$, $QQ'$, $RR'$, $SS'$ all pass through $O$. Since $P$, $P'$, $S$, $S'$ all lie in one plane $OPS$, the lines $PS$, $P'S'$ meet in a point $P_1$ (possibly at infinity); similarly $QS$ meets $Q'S'$ in a point $Q_1$, and $RS$ meets $R'S'$ in a point $R_1$. Applying the "obvious" version of the theorem to the triangles $QRS$, $Q'R'S'$, which lie in distinct planes, we see that the points of intersection

$$R_1 = RS \cdot R'S', \qquad Q_1 = SQ \cdot S'Q', \qquad A = QR \cdot Q'R'$$

are collinear. Thus $A$ lies on $Q_1R_1$; similarly $B$ on $R_1P_1$, and $C$ on $P_1Q_1$. Hence the three points $A$, $B$, $C$, lying in the plane $P_1Q_1R_1$ as well as in the plane $PQR$, must lie on the line of intersection $PQR \cdot P_1Q_1R_1$.

Girard Desargues (1593–1662) was an architect of Lyons. He discovered not only the above theorem but also several others, which we shall use later, especially in connection with conics. His treatise of 1639 was not well received during his lifetime, partly because of his obscure style: he introduced about seventy new terms, of which only *involution* has survived.

EXERCISE

By taking the corresponding vertices $R$ and $R'$ to be points at infinity, deduce from 1·52 the converse of 1·51: *If two triangles have parallel sides, the joins of their corresponding vertices are concurrent or parallel.*

**1·6   An Outline of Subsequent Work.**   In the next chapter we shall make a fresh start, considering real projective geometry as a self-contained system, defined by its own peculiar axioms. We have the satisfaction of knowing that it is *consistent* (*i.e.*, that its axioms cannot lead to any contradictory statements) because we can obtain a "model" of it by adding the line at infinity to the affine plane. Here we are taking for granted the consistency of Euclidean geometry, which includes affine geometry.

This investigation of projective geometry will be continued throughout Chaps. 2 to 7. Then we shall derive affine geometry in Chap. 8 and Euclidean in Chap. 9. At that stage we shall have returned to our starting point, ready to deal in a more sophisticated manner with the difficult subject of continuity (Chap. 10). Finally, in Chaps. 11 and 12 we shall see how these *synthetic* geometries lead to, and can be derived from, *analytic* geometry.

**1·7   The Directed Angle, or Cross.*** One of the concepts to which we shall be led in Chap. 9 is that of *angle:* not the customary "angle between two rays" but an "angle between two *lines*," which is subtly different. From this standpoint, $\langle AOB \rangle$ means the angle through which a variable line has to be turned, in the counterclockwise sense, in order to pass from the position $AO$ to the position $OB$. Thus $\langle AOB \rangle$ is not altered by shifting $A$ along $AO$, or $B$ along $OB$, even beyond $O$. Such angles may be measured in degrees or radians, but they cannot be negative and are always less than 180° or $\pi$. The angles $\langle AOB \rangle$ and $\langle BOA \rangle$ are supplementary, as we see in Fig. 1·7A. Here are some of Euclid's propositions expressed in terms of these directed angles:

* Johnson (Ref. 22, pp. 12–15); Forder (Ref. 12, p. 120). Picken's article, Euclidean Geometry of Angle, appeared in the *Proceedings of the London Mathematical Society* (2), vol. 23, pp. 45–55, 1925. See also Forder (Ref. 13) and his article, The Cross and the Foundations of Euclidean Geometry, *Mathematical Gazette*, vol. 31, pp. 227–233, 1947.

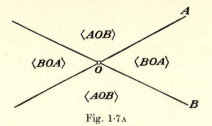

Fig. 1·7ᴀ

I. 5. *If triangle ABC has equal sides AB and AC, then*

$$\langle ABC \rangle = \langle BCA \rangle.$$

I. 13. $\langle AOB \rangle + \langle BOA \rangle = \pi.$

I. 27. *If* $\langle PAB \rangle = \langle QBA \rangle$, *then AP is parallel to BQ.*

III. 21 and 22. *If A, B, C, D lie on a circle, then* $\langle ACB \rangle = \langle ADB \rangle.$

III. 32. *If AT is the tangent at A to the circle ABC, then*

$$\langle ACB \rangle = \langle TAB \rangle.$$

This kind of angle was invented independently by R. A. Johnson in America and D. K. Picken in New Zealand.

### EXERCISES

1. Show that $\langle ACD \rangle = \langle BCD \rangle$ if A, B, C are collinear.

2. Show that $\langle AOB \rangle = \langle BOA \rangle$ if OA and OB are perpendicular.

3. Show that $\langle AOX \rangle = \langle XOB \rangle$ if OX is the internal or external bisector of the angle $\langle AOB \rangle$.

# CHAPTER 2

# INCIDENCE

The geometry considered in this book is called *real* because, if we chose to work it out analytically, the coordinates would be *real* numbers, whereas otherwise they might have been complex numbers, or the "numbers" of a finite arithmetic (Galois field),* or something still more bizarre. However, the present chapter deals with those properties of the projective plane which depend only on the simple processes of joining and intersection and which are consequently valid in the other geometries mentioned above, as well as in real geometry. These properties include the principle of duality, perspectivity, and harmonic conjugacy. Many of the ideas can be traced back to Desargues (who defined harmonic conjugates by dividing a segment internally and externally in the same ratio), but their essentially projective nature was first understood by an extraordinarily talented German, von Staudt (1798–1867).

**2·1  Primitive Concepts.**  In a logical development of geometry, each definition of an entity or relation involves other entities or relations; therefore some entities and relations, the *primitive concepts*, must remain undefined. Similarly, the proof of each proposition uses other propositions; therefore some propositions, the *axioms*, must remain unproved. In practice, the primitive concepts should have some intuitive significance, some interpretation in which the axioms are seen to be true. Otherwise we should be playing a meaningless game.

A basis for the system of real projective geometry may be chosen in many different ways. It seems simplest to take as primitive concepts *point, line, incidence,* and *separation*.

There is no harm in picturing a point as the idealized limit of smaller and smaller material dots ("position without magnitude") and a line as the idealized limit of a material line drawn with a sharp pencil on smooth paper against a straight ruler. Of course, a microscope would

---

* Robinson (Ref. 32, pp. 87–89, 106–108).

reveal imperfections in any material line, but the geometrical object is supposed to be perfectly thin and straight and infinitely extended.

A point and a line may or may not be *incident*. When they are, we say that the point lies on the line or that the line passes through the point. A line passing through two points is called their *join*, and a point lying on two lines is called their *intersection*.

The relation of *separation* applies to two pairs of points on a line or to two pairs of lines through a point. If four such points (or lines) $A$, $B$, $C$, $D$ occur in that order, we say that $A$ and $C$ *separate* $B$ and $D$ and write for brevity

$$AC//BD$$

This relation has a less straightforward meaning in complex geometry and no meaning whatever in finite geometries, but it does properly

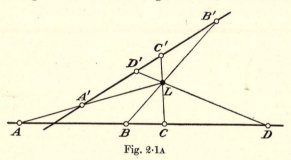

Fig. 2·1A

belong to *real* projective geometry, for it is invariant under central projection (see Fig. 2·1A, where $AC//BD$ and $A'C'//B'D'$.) On the other hand, the simpler notion of $B$ lying "between" $A$ and $C$, which belongs to affine geometry, is not invariant. (Visibly, $B'$ is not between $A'$ and $C'$.) In fact, the kind of order that belongs to real projective geometry is not serial but cyclic. We shall return to these considerations in the next chapter.

Although the above description of the primitive concepts helps our imagination and thus suggests what axioms are appropriate, we must take care never to use any of these intuitive ideas in our proofs. The only properties to be assumed are those actually stated in the axioms.

The two processes of joining and intersection, which emerge from the relation of incidence, somewhat resemble the processes of addition and multiplication in algebra and are sometimes denoted by the same symbolism. We shall adopt the "multiplicative" symbol $a \cdot b$ for the intersection of lines $a$ and $b$, but the join of points $A$ and $B$ will be denoted by the familiar symbol $AB$, rather than the more startling

$A + B$. These symbols are easy to combine; *e.g.*, the intersection of $AB$ and $CD$ is $AB \cdot CD$, while the join of $a \cdot b$ and $c \cdot d$ is $(a \cdot b)(c \cdot d)$.

### 2·2   The Axioms of Incidence.

**2·21**   *There exist a point and a line that are not incident.*

**2·22**   *Every line is incident with at least three points.*

**2·23**   *Any two points are incident with just one line.*

**2·24**   *Any two lines are incident with at least one point.*

**2·25**   *If the three lines $PP'$, $QQ'$, $RR'$ are all incident with one point, then the three points $QR \cdot Q'R'$, $RP \cdot R'P'$, $PQ \cdot P'Q'$ are all incident with one line.*

Such a set of axioms (using point, line, and incidence as primitive concepts) was given in 1899.\* They are all very simple except the fifth, which we are prepared to accept because it is the same as Desargues' two-triangle theorem (our theorem 1·52). (It cannot be deduced from the four simple axioms; for there exist "non-Desarguesian" geometries† that satisfy 2·21 to 2·24 without satisfying 2·25.) We proceed to prove its converse:

**2·26**   *If two triangles have corresponding sides intersecting in collinear points, then the joins of corresponding vertices are concurrent.*

*Proof:* Using the same notation as in 1·52, we have two triangles $PQR$ and $P'Q'R'$ whose corresponding sides intersect in the three collinear points $A$, $B$, $C$, and we wish to prove that the line $RR'$ passes through the point $O = PP' \cdot QQ'$ (see Fig. 2·2A). This is an immediate consequence of Desargues' two-triangle theorem itself, as applied to the triangles $AQQ'$ and $BPP'$, whose joins of corresponding vertices all pass through $C$, while their intersections of corresponding sides are $O$, $R'$, $R$.

### EXERCISES

1. Give detailed proofs of the following simple theorems, pointing out which axioms are used:

    *a.* Every point is incident with at least three lines.

    *b.* Any two lines are incident with just one point.

2. The two triangles $PQR$ and $P'Q'R'$ of Fig. 2·2A are conveniently said to be *in perspective* from the *center* $O$ and *axis* $ABC$. If three triangles are all in perspective from the same center, prove that the three axes are concurrent. (*Hint:* Let the three axes be $A_1B_1$, $A_2B_2$, $A_3B_3$. Apply 2·26 to triangles $A_1A_2A_3$ and $B_1B_2B_3$.)

* Pieri (Ref. 28, pp. 6–22, Postulates I–XIII).

† See, *e.g.*, Robinson (Ref. 32, pp. 126–128). Such geometrics can occur only in two dimensions. Pieri avoided the assumption of 2·25 by working in three dimensions.

3. Show that the 10 points and 10 lines of the Desargues configuration (Fig. 2·2A or B) may be renamed $P_{12}$, . . . , $P_{45}$ and $p_{12}$, . . . , $p_{45}$ in such a way that $P_{ij}$ and $p_{kl}$ are incident whenever the numbers represented by $i$, $j$, $k$, $l$ are all different. (*Hint:* Let $PQRP'Q'R'$ be called $P_{14}P_{24}P_{34}P_{15}P_{25}P_{35}$. This notation arises from the figure of 5 points in space with the 10 lines and 10 planes that join them. Taking the section by a plane of general position, we obtain $P_{12}$ and $p_{12}$ as sections of the line 12 and plane 345. The five points 1, 2, 3, 4, 5 may be identified with the $P_1$, $Q_1$, $R_1$, $S$, $S'$ of Sec. 1·5.)

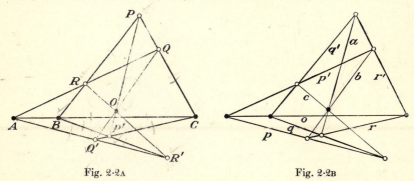

Fig. 2·2A                              Fig. 2·2B

4. Show that the same 10 points and 10 lines may be regarded (in six ways) as consisting of two pentagons so situated that consecutive sides of each pass through alternate vertices of the other. (*Hint:* Consider the pentagons $P_{12}P_{23}P_{34}P_{45}P_{51}$ and $P_{31}P_{14}P_{42}P_{25}P_{53}$ of Exercise 3.)

**2·3  The Principle of Duality.** The principle of duality (in two dimensions) asserts that every definition remains significant, and every theorem remains true, when we interchange the two pairs of concepts:

*point* and *line*,

*join* and *intersection*.

Thus the dual of $AB \cdot CD$ is $(a \cdot b)(c \cdot d)$, Axiom 2·21 is self-dual, and the dual of 2·24 is part of 2·23. Some other changes of wording are obvious consequences of these fundamental changes; *e.g.*, the dual of 1·52 is 2·26.

To establish this principle we merely have to observe that *the axioms imply their own duals* (see Sec. 2·2, Exercise 1). Given a theorem and its proof, we can immediately assert the dual theorem; for a proof of the latter could be written down mechanically by dualizing every step in the proof of the original theorem.

Although the closely related idea of reciprocal polyhedra had already occurred in the writings of the medieval Italian Maurolycus (1494–1575), the principle of duality may properly be ascribed to Gergonne (1771–1859).

Poncelet protested that it was nothing but his method of reciprocation with respect to a conic (polarity), and Gergonne replied that the conic is irrelevant —duality is intrinsic in the system. Thus Gergonne came nearer to realizing how the principle rests on the symmetrical nature of the axioms of incidence. It is sad that such a beautiful discovery was marred by bitter controversy over the question of priority.

Instead of using axioms that merely *imply* their duals, it might perhaps be more satisfactory to use an inherently self-dual set of axioms, such as the following:*

**2·31**   *Two distinct $\begin{smallmatrix}points\\lines\end{smallmatrix}$ are incident with at least one $\begin{smallmatrix}line\\point\end{smallmatrix}$.*

**2·32**   *Two distinct points cannot both be incident with two distinct lines.*

**2·33**   *There exist two points and two lines such that each of the points is incident with just one of the lines.*

**2·34**   *There exist two points and two lines (the points not incident with the lines) such that the join of the points is incident with the intersection of the lines.*

**2·35**   *If four points O, P, Q, R having six distinct joins and four lines o, p, q, r having six distinct intersections are so situated that the five joins OP, OQ, OR, PR, QR are incident with the respective intersections $q \cdot r$, $r \cdot p$, $p \cdot q$, $q \cdot o$, $o \cdot p$, then the sixth join PQ is incident with the sixth intersection $o \cdot r$.*

### EXERCISES

1. Express 2·26 in terms of lines $p$, $q$, $r$, $p'$, $q'$, $r'$, so as to make it formally dual to 2·25 (see Fig. 2·2B).

2. Verify that Axioms 2·21 to 2·25 imply 2·31 to 2·35, and vice versa.

**2·4   Quadrangle and Quadrilateral.**   The following definitions are written in parallel columns to emphasize the principle of duality:

Four points $P$, $Q$, $R$, $S$, of which no three are collinear, are the vertices of a complete *quadrangle*† *PQRS*, of which the six sides are the lines $QR$, $PS$, $RP$, $QS$, $PQ$, $RS$. The intersections of "opposite" sides, namely,

$$A = QR \cdot PS$$
$$B = RP \cdot QS$$
$$C = PQ \cdot RS$$

are called *diagonal points* and are the vertices of the *diagonal triangle* (see Fig. 2·4A).

Four lines $p$, $q$, $r$, $s$, of which no three are concurrent, are the sides of a complete *quadrilateral*† *pqrs*, of which the six vertices are the points $q \cdot r$, $p \cdot s$, $r \cdot p$, $q \cdot s$, $p \cdot q$, $r \cdot s$. The joins of "opposite" vertices, namely,

$$a = (q \cdot r)(p \cdot s)$$
$$b = (r \cdot p)(q \cdot s)$$
$$c = (p \cdot q)(r \cdot s)$$

are called *diagonal lines* and are the sides of the *diagonal triangle* (see Fig. 2·4B).

* Axioms 2·31 to 2·34 were kindly supplied by Karl Menger. For 2·35, see Veblen and Young (Ref. 42, p. 53, Exercise 8).

† When there is no danger of confusion, we shall omit the word *complete*.

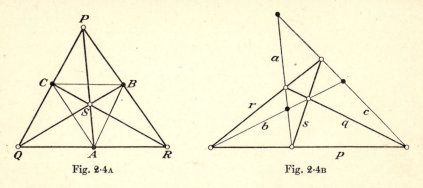

Fig. 2·4A          Fig. 2·4B

**2·41**  *If ABC is the diagonal triangle of a quadrangle PQRS, the three points*

$$A_1 = BC \cdot QR, \qquad B_1 = CA \cdot RP, \qquad C_1 = AB \cdot PQ$$

*are collinear.*

*Proof:* Apply Desargues' theorem (our theorem 1·52 or Axiom 2·25) to the two triangles $ABC$ and $PQR$ (see Fig. 2·4c).

Fig. 2·4c

As a corollary we have

**2·42**  *Given the diagonal triangle and one vertex of a quadrangle, the remaining three vertices may be constructed by incidences.*

In fact, given the diagonal triangle $ABC$ and vertex $P$, we construct, in turn,

$$B_1 = CA \cdot BP, \qquad C_1 = AB \cdot CP, \qquad A_1 = BC \cdot B_1C_1$$
$$R = BP \cdot AA_1, \qquad Q = CP \cdot AA_1, \qquad S = AP \cdot BQ$$

### EXERCISES

1. Prove that the six sides of a quadrangle meet the three sides of its diagonal triangle in the six vertices of a quadrilateral which has the same diagonal tri-

angle. *Hint:* Define

$$A_2 = BC \cdot PS, \qquad B_2 = CA \cdot QS, \qquad C_2 = AB \cdot RS$$

and use triangles $ABC$ and $SRQ$ to prove $A_1$, $B_2$, $C_2$ collinear.

2. Dualize 2·41 and 2·42.

3. Show that the 10 points and 10 lines of the Desargues configuration (Fig. 2·2A or B) may be regarded (in five ways) as consisting of a quadrangle and a quadrilateral so situated that the six sides of the quadrangle pass through the six vertices of the quadrilateral (cf. 2·35).

**2·5 Harmonic Conjugacy.** Although harmonic conjugates were used by Desargues, the following construction for them seems to have been first given by another Frenchman, La Hire (1640–1718).

Four collinear points $A$, $B$, $C$, $D$ are said to form a *harmonic set* if there is a quadrangle of which two opposite sides pass through $A$ and two other opposite sides through $B$, while the remaining sides pass through $C$ and $D$, respectively. We say that $C$ and $D$ are *harmonic conjugates* (of each other) wo* $A$ and $B$, and we write

$$\text{H}\,(AB,\ CD)$$

as an abbreviated statement of this relation.

To construct $D$, given $A$, $B$, $C$, we draw any triangle $PQR$ whose sides $QR$, $RP$, $PQ$ pass through $A$, $B$, $C$, respectively. This determines a quadrangle $PQRS$, where

$$S = AP \cdot BQ,$$

as in Fig. 2·5A. We thus obtain

$$D = RS \cdot AB$$

Four concurrent lines $a$, $b$, $c$, $d$ are said to form a *harmonic set* if there is a quadrilateral of which two opposite vertices lie on $a$ and two other opposite vertices on $b$, while the remaining vertices lie on $c$ and $d$, respectively. We say that $c$ and $d$ are *harmonic conjugates* (of each other) wo* $a$ and $b$, and we write

$$\text{H}(ab,\ cd)$$

as an abbreviated statement of this relation.

To construct $d$, given $a,b,c$, we draw any triangle $pqr$ whose vertices $q \cdot r$, $r \cdot p$, $p \cdot q$ lie on $a$, $b$, $c$, respectively. This determines a quadrilateral $pqrs$, where

$$s = (a \cdot p)(b \cdot q),$$

as in Fig. 2·5B. We thus obtain

$$d = (r \cdot s)(a \cdot b)$$

Fig. 2·5A

Fig. 2·5B

* The preposition *wo* (pronounced like "woe") has been coined by some English mathematicians as a convenient abbreviation for "with respect to" or "with regard to."

The construction for $D$ or $d$ involves the choice of a triangle. How do we know that a different triangle will lead to the same final result? This question of the *uniqueness* of the harmonic conjugate can be answered affirmatively with the help of Sec. 2·2. We need only consider the case of a harmonic set of points; that of a harmonic set of lines will follow by duality (because we know that the proof could be dualized step by step).

**2·51**   *The harmonic conjugate of C wo A and B is independent of the choice of triangle PQR.*

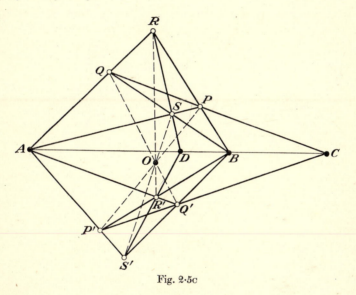

Fig. 2·5c

*Proof:* Suppose that another such triangle $P'Q'R'$ leads to a quadrangle $P'Q'R'S'$, as in Fig. 2·5c. We have to show that $RS$ and $R'S'$ both determine the same point $D$ on $AB$. For this purpose we consider, in turn, three pairs of triangles. Corresponding sides of triangles $PQR$ and $P'Q'R'$ meet in the three collinear points $A$, $B$, $C$; hence, by 2·26, the joins of corresponding vertices are concurrent, that is, $RR'$ passes through the point $O = PP' \cdot QQ'$. Applying the same theorem to triangles $PQS$ and $P'Q'S'$, we conclude that $SS'$ passes through this same point $O$. Thus the joins of corresponding vertices of triangles $RSP$ and $R'S'P'$ all pass through $O$; hence, by Desargues' theorem, their corresponding sides meet in collinear points. But two of these points are $A$ and $B$; therefore the remaining sides $RS$ and $R'S'$ both meet $AB$ in the same point $D$.

In the notation of Fig. 2·4c, we have H($AB$, $C_1C_2$). Thus the sides of the diagonal triangle are "divided harmonically" by the sides of the quadrangle. Moreover,

**2·52** *The sides of a quadrangle are divided harmonically by the sides of its diagonal triangle.*

*Proof:* The quadrangle $PSBC$ yields H($QR$, $AA_1$).

Incidentally, we observe that the quadrangle $CC_1BB_1$ yields H($AA_1$, $QR$), which is not obviously the same as H($QR$, $AA_1$).

### EXERCISES

1. Using a pencil and ruler, carry out the construction for the harmonic conjugate of $C$ wo $A$ and $B$, taking $C$ somewhere between $A$ and $B$. What happens if $C$ is midway between $A$ and $B$?

2. Dualize 2·51 and its proof, drawing a suitable figure.

**2·6  Ranges and Pencils.** The points on a line are said to form a *range*, especially when we regard them as the possible positions of a

Fig. 2·6A

*variable* point $X$ (which "runs along" the line). The dual of a range is a *pencil*, consisting of the lines through one point: the possible positions of a variable line $x$ (which "rotates about" the point). The common point of the lines is called the *center* of the pencil.

We proceed to define a *correspondence* (strictly, a one-to-one correspondence) between two ranges. This is a rule for associating every point $X$ of the first range with every point $X'$ of the second, so that there is exactly one $X'$ for each $X$ and exactly one $X$ for each $X'$. It is usually desirable to think of the correspondence as being directed from $X$ to $X'$, that is, to distinguish between this and the *inverse* correspondence from $X'$ to $X$. The two ranges need not be on distinct lines. One trivial case, which must not be ignored, is when $X'$ continually coincides with $X$; this correspondence is called the *identity*.

There is a similar definition for a correspondence between two pencils or between a pencil and a range. The simplest correspondence of the latter type occurs when we take the section of the pencil by a fixed line $o$, so that each line $x$ of the pencil is associated with the point

$X = o \cdot x$ of the corresponding range, as in Fig. 2·6A. The inverse correspondence occurs when we project a range from a fixed point $O$ so that each point $X$ of the range is associated with the line $x = OX$ of the corresponding pencil. The existence of these simple correspondences is one of the basic reasons for the efficacy of projective geometry.

In affine (or Euclidean) geometry, the line $p$ (through $O$) parallel to $o$ would be exceptional: it would have no corresponding point on $o$. But when we have extended the affine plane to the projective plane, the corresponding point $P$ is just the point at infinity on $o$. The line $x$ through $O$, rotating continuously, determines on $o$ the point $X$, which runs along to the right, say, until $x$ is parallel to $o$, then immediately reappears far away on the left and continues running to the right. In affine geometry the point $X$ makes an infinite jump; but in projective geometry its motion, through the single point at infinity, is continuous.

**2·7  Perspectivity.** Apart from the identity, the simplest correspondence between two ranges is that which occurs when we compare

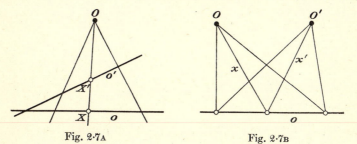

Fig. 2·7A                    Fig. 2·7B

the sections of a pencil by two distinct lines $o$ and $o'$, as in Fig. 2·7A. The relation between $X$ on $o$ and $X'$ on $o'$ is such that the line $XX'$ passes through a fixed point $O$, and we call the correspondence a *perspectivity* from $O$, writing

$$X \overset{O}{\barwedge} X', \qquad \text{or simply} \qquad X \barwedge X'$$

Dually (Fig. 2·7B) a perspectivity from a line $o$ occurs when the relation between two pencils is such that the point of intersection $x \cdot x'$ lies on a fixed line $o$; then we write

$$x \overset{o}{\barwedge} x' \qquad \text{or} \qquad x \barwedge x'$$

The following important theorem illustrates the way this notation may be used:*

---

* von Staudt (Ref. 40, p. 59, Sec. 119). This use of the symbol $\barwedge$ is due to Veblen and Young (Ref. 42, p. 57). (von Staudt used it differently.)

**2·71** *It is possible, by a sequence of three perspectivities, to inter-change pairs among any four collinear points.*

*Proof:* Suppose we wish to interchange $A$ with $A'$, and $B$ with $B'$, that is, to make the permutation $AA'BB' \to A'AB'B$ or $(AA')(BB')$. Draw any triangle $PQR$ whose sides $QR, RP, PQ$ pass through $A, B, B'$. This determines two further points,

$$U = A'R \cdot PQ, \qquad V = AU \cdot RP$$

as in Fig. 2·7c, and we have

$$AA'BB' \; \overset{R}{\overline{\wedge}} \; QUPB' \; \overset{A}{\overline{\wedge}} \; RVPB \; \overset{U}{\overline{\wedge}} \; A'AB'B$$

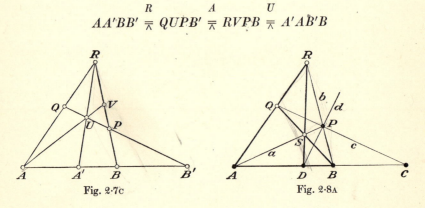

Fig. 2·7c            Fig. 2·8A

**2·8 The Invariance and Symmetry of the Harmonic Relation.** We proceed to show that harmonic sets remain harmonic after any number of perspectivities. As a first step we shall prove

**2·81** *Any section of a harmonic set of lines is a harmonic set of points, and a harmonic set of points is projected from any point by a harmonic set of lines.*

*Remark:* This theorem is in two dual parts, and thus it will suffice to prove the latter part: If $A$, $B$, $C$, $D$ are joined to a point $P$ (outside their line) by lines $a$, $b$, $c$, $d$ and if H($AB$, $CD$), then H($ab$, $cd$).

*Proof:* Let $P$ be used as a vertex of the triangle $PQR$ in constructing $D$ from $A$, $B$, $C$, as in Fig. 2·5A or 2·8A. Then the quadrilateral $ASBRQD$ has two opposite vertices on $AS = a$, two others on $BR = b$, one vertex $Q$ on $c$, and one vertex $D$ on $d$. Hence H($ab$, $cd$).

As a corollary we have

**2·82** *Perspectivities preserve the harmonic relation:*
If $ABCD \; \overset{}{\overline{\wedge}} \; A'B'C'D'$ and H($AB$, $CD$), then H($A'B'$, $C'D'$).

Combining this result with 2·71, we infer that H($AB$, $CD$) implies H($CD$, $AB$). But our definition for harmonic conjugacy involves $A$ and $B$ symmetrically and likewise $C$ and $D$. Hence

**2·83**　*The eight relations*

|  |  |  |  |
|---|---|---|---|
| H($AB$, $CD$), | H($BA$, $CD$), | H($AB$, $DC$), | H($BA$, $DC$), |
| H($CD$, $AB$), | H($CD$, $BA$), | H($DC$, $AB$), | H($DC$, $BA$) |

*are all equivalent.*

### EXERCISES

1. Show that Fig. 2·8A can be made formally self-dual by interchanging the names of $c$ and $d$ (so that $c = PD$ and $d = PC$). Which lines should then be named $p$, $q$, $r$, $s$?

2. If we have H($AB$, $CD$) and H($A'B'$, $C'D$) on distinct lines, show that the three lines $AA'$, $BB'$, $CC'$ are concurrent.

# CHAPTER 3

# ORDER AND CONTINUITY

The order of arrangement of lines in a pencil, like that of points on a circle, is cyclic: we cannot say of three that one is between the other two, but we can say of four that two *separate* the other two. The correspondence between a pencil and its section enables us to carry over this cyclic order from pencils to ranges. If $A$ and $B$ separate $C$ and $D$, we write $AB//CD$. (The idea of a point $C$ lying *between* $A$ and $B$ belongs to affine geometry and may be interpreted as meaning that $AB//CD$ where $D$ is the point at infinity on $AB$.)

The basic properties of separation may be stated in the form of six axioms, as in Sec. 3·1. These are not quite sufficient for a complete characterization of the real projective line. The final axiom, concerning continuity, will be introduced in Sec. 3·5. This particular form has been chosen because it is ready for immediate application in proving the fundamental theorem of projective geometry and other theorems; moreover, it is intuitively acceptable. Dedekind's axiom has not been used, for it is more difficult to grasp and to apply. The deduction of our axiom from Dedekind's was carried out by Enriques.[*] In Chap. 10 we shall consider a third possible approach to the theory of continuity.

## 3·1 The Axioms of Order.

**3.11** *If $A$, $B$, $C$ are three distinct collinear points, there is at least one point $D$ such that $AB//CD$.*

**3·12** *If $AB//CD$, then $A$, $B$, $C$, $D$ are distinct.*

**3·13** *If $AB//CD$, then $AB//DC$.*

**3·14** *If $A$, $B$, $C$, $D$ are four distinct collinear points, at least one of the three relations $BC//AD$, $CA//BD$, $AB//CD$ must hold.*

**3·15** *If $AB//CD$ and $AC//BE$, then $AB//DE$.*

**3·16** *If $AB//CD$ and $ABCD \mathbin{\overline{\overline{\wedge}}} A'B'C'D'$, then $A'B'//C'D'$.*

The first five of these six axioms have been adapted from those given by Vailati. They express obvious properties of points arranged

[*] Enriques (Ref. 11, pp. 71–75). Cf. Sec. 10·6.

round a circle. Accordingly, we often find it convenient to use a circular diagram when dealing with points on a single line; *e.g.*, Fig. 3·1A illustrates 3·15. (This does not mean that we imagine the line to be somehow "bent," but it emphasizes the important fact that the line is "closed.")

In contrast to these five one-dimensional axioms, 3·16 is essentially two-dimensional, relating order on one line to order on another. It enables us to derive the dual statements concerning the relation $ab//cd$ for concurrent lines.

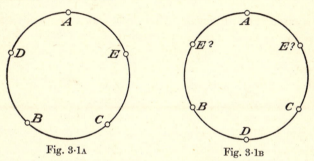

Fig. 3·1A                          Fig. 3·1B

Using 2·71 to interchange the pairs $(AC)(BD)$, we deduce from 3·16 that

$$AB//CD \qquad \text{implies} \qquad CD//AB$$

Taking this with 3·13, we conclude that the eight relations

$$AB//CD, \qquad BA//CD, \qquad AB//DC, \qquad BA//DC$$
$$CD//AB, \qquad CD//BA, \qquad DC//AB, \qquad DC//BA$$

are all equivalent (cf. 2·83).

By 3·12, the $D$ and $E$ of 3·15 cannot coincide. Thus the relation $AB//CD$ excludes $AC//BD$. But these are equivalent to two of the relations in 3·14, and we can argue similarly for any other two. Hence

**3·17**   *The three relations $BC//AD$, $CA//BD$, $AB//CD$ are mutually exclusive: no two can hold simultaneously.*

The following theorem has a somewhat similar appearance:

**3·18**   *The three relations $BC//DE$, $CA//DE$, $AB//DE$ cannot all hold simultaneously.*

*Proof:* Assuming all three, we shall find that each of the relations in 3·14 leads to a contradiction. Suppose, for instance, that $BC//AD$ (or $AD//BC$), as in Fig. 3·1B. Then by 3·15 the relations $AB//DE$ and $AD//BC$ imply $AB//EC$, while $AC//DE$ and $AD//CB$ imply $AC//EB$ (or $CA//BE$). But the conclusions $CA//BE$ and $AB//CE$ are incompatible, by 3·17. Since $A$, $B$, $C$ enter the discussion symmetrically, the proof is complete.

This tricky proof of an obvious statement indicates the economy of our axioms: they are only just strong enough to provide the familiar properties of cyclic order.

### EXERCISE

Prove that $AB//CD$ and $AC//BE$ imply $BE//CD$.   (*Hint:* $BE//AC$ and $BA//ED$.)

**3·2   Segment and Interval.**   If $A$, $B$, $C$ are three collinear points, we define the *segment* $AB/C^*$ as consisting of all points $X$ for which $AB//CX$. (Thus the segment $AB/C$ does *not* contain $C$. The familiar "segment $AB$" of affine geometry may be described as $AB/C$ with $C$ at infinity.) The segment plus its end points $A$ and $B$ is called an *interval* and written $\overline{AB}/C$. If $X$ and $Y$ belong to $\overline{AB}/C$, the interval $\overline{XY}/C$ is said to be *interior* to $\overline{AB}/C$ (even if $X$ or $Y$ coincides with $A$ or $B$), and a point $D$ lies *between* $X$ and $Y$ in $\overline{AB}/C$ if it belongs to $XY/C$, that is, if $XY//CD$. Thus the notion of intermediacy (or three-point order) is valid for an interval, although not for the whole line. In a circular diagram such as Fig. 3·1A an interval naturally appears as an arc.

By drawing several diagrams like Fig. 2·5A the reader will easily convince himself that the harmonic conjugate of $C$ wo $A$ and $B$ lies in the segment $AB/C$. But to prove this rigorously is not so easy. The following proof is due to Enriques:[†]

**3·21**   If $A$, $B$, $C$ *are all distinct,* $\mathrm{H}(AB, CD)$ *implies* $AB//CD$.

*Proof:* By 3·11 we can take a point $M$ such that $AS//PM$, as in Fig. 3·2A. Let $QM$ meet $AB$ in $Y$, and $RS$ in $O$. Let $PO$ meet $AB$ in $X$, and $AR$ in $N$. If $Y$ happens to coincide with $D$, we immediately obtain $AB//CD$ by perspectivity from $Q$. If not, we have

$$ASPM \overset{O}{\barwedge} ARNQ \overset{P}{\barwedge} ABXC, \quad ASPM \overset{Q}{\barwedge} ABCY, \quad ASPM \overset{O}{\barwedge} ADXY$$

Hence by 3·16 we have $AB//XC$, $AB//CY$, $AD//XY$. Thus both $X$ and $Y$ are in the segment $AB/C$, and $D$ is successfully trapped between them.[‡]

---

* Read as "$AB$ without $C$."

† Ref. 11, p. 51.

‡ The details involved in this "trapping" are as follows: By 3·18 the relations $AB//XC$ and $AB//CY$ exclude $AB//XY$; thus, by 3·14, we must have either $AX//BY$ or $AY//BX$. We may assume the former possibility (as the latter can be treated by the consistent

**3·22**　Corollary: *If A, B, C are all distinct,* H(*AB, CD*) *implies* *D ≠ C.*

In other words, *the diagonal points of a quadrangle are not collinear.* (Consider the quadrangle *ABPQ.*) This result, which we tacitly assumed in Sec. 2·5, is sometimes taken as an axiom.*

**3·3**　**Sense.**　In ordinary (affine) geometry a point decomposes a line through it into two rays; but in projective geometry a point does not divide a line at all. (We can reach the left side of the barrier point

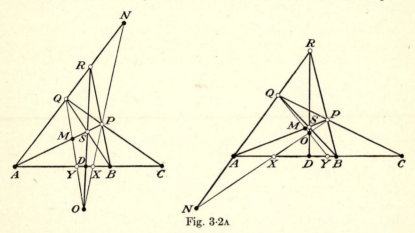

Fig. 3·2A

from the right by proceeding to the right and passing through the point at infinity.) The circular diagram suggests that two points will decompose their joining line into just two segments (represented by two semicircles, say); but this is quite difficult to prove, as we shall see.

**3·31**　*If AB//CD, the two points A and B decompose their line into just two segments: AB/C and AB/D.*

*Proof:* No point *X* can lie in both *AB/C* and *AB/D*; for then we should have *XC//AB, CD//AB, DX//AB,* which are incompatible (by 3·18). It remains to be shown that any point *X*, other than *A* or *B*, must lie in *one* of the segments.† We shall assume that *X* does not lie in *AB/C* and deduce that it then lies in *AB/D*. By 3·14 (with *X* for *D*) either *BC//AX* or *CA//BX*. In the former case we have *BA//CD*

---

interchange of *X* and *Y*). By 3·15 the relations *XA//YB* and *XY//AD* imply *XA//BD*; finally, *AB//XC* and *AX//BD* imply *AB//CD*.

* Veblen and Young (Ref. 42, p. 45).

† This part of the proof is due to Robinson (Ref. 32, p. 120).

and $BC//AX$, implying $BA//DX$; in the latter, $AB//CD$ and $AC//BX$, implying $AB//DX$. Thus in either case $X$ belongs to $AB/D$, as required.

Two such segments, and likewise the corresponding intervals $\overline{AB}/C$ and $\overline{AB}/D$, are said to be *supplementary*.

Axiom 3·14 and Theorem 3·17 imply that three collinear points $A$, $B$, $C$ decompose their line into three segments $BC/A$, $CA/B$, $AB/C$; and it can be proved by induction that $n$ collinear points decompose their line into $n$ segments.* Thus the line contains infinitely many points.

In this manner the notion of the cyclic order of any number of points may be rigorously justified, but it would be tedious to give all the logical details here. Enough has been said to show the reader that he can henceforth safely rely on intuition, in the certainty that he could, if he wished, supply the proofs.

One consequence of this notion of cyclic order is the distinction of *sense*.† If $D$, $E$, $F$ are three distinct points on the line $ABC$, we can say whether the sense $DEF$ agrees or disagrees with the sense $ABC$, writing, respectively,

$$S(ABC) = S(DEF) \qquad \text{or} \qquad S(ABC) \neq S(DEF)$$

Any of $D$, $E$, $F$ may coincide with any of $A$, $B$, $C$. In particular,

$$S(ABC) = S(BCA) = S(CAB) \neq S(CBA)$$

Instead of deriving sense from separation, Veblen considered "undefined elements called senses," in terms of which he defined separation. In the present treatment his definition becomes a theorem:

**3·32** *The relation $AB//CD$ is equivalent to* $S(ABC) \neq S(ABD)$.

For brevity, the formal proof is omitted.† In Fig. 2·5A the sense $S(ABC)$ is "left to right," while $S(ABD)$ or $S(BDA)$ is "right to left." In Fig. 3·1A, $S(ABC)$ is "positive," or "counterclockwise," while $S(ABD)$ is "negative," or "clockwise."

The following theorem is not needed for the subsequent development, but is interesting because, after its enunciation by Sylvester in 1893, it remained unproved for about forty years. Then T. Grünwald proved it by an ingenious argument using parallel lines. The projective proof given here is due to R. Steinberg.‡

---

* Veblen and Young (Ref. 43, p. 46).

† Coxeter (Ref. 6, p. 32). Cf. Veblen and Young (Ref. 43, p. 32).

‡ *American Mathematical Monthly*, vol. 51, pp. 169–171, 1944; vol. 55, pp. 26–28, 247, 1948; *Mathematical Reviews*, vol. 9, p. 458, 1948.

**3·33** *Let n given points have the property that the line joining any two of them passes through a third point of the set. Then the n points are all collinear.*

*Proof:* We shall suppose three of the $n$ points to form a triangle $PQR$ and shall show that this leads to a contradiction. Let $p$ be a line through $P$ that contains no other point of the set. All the joins of pairs of the $n$ points meet $p$ in a certain set of at least two points: $P$ itself, one on $QR$, and possibly others. These points occur in a certain cyclic order. Let $A$ be consecutive to $P$ in this order, so that one of the segments $AP$ is not met by any of the joins. This point $A$ is not one of the $n$ but lies on a line containing at least three of them, say $B$, $C$, $D$, so named that $AB//CD$. Since $P$ and $B$ are two of the $n$ points, their join must contain a third, say $O$. Suppose $ABCD \overset{O}{\overline{\overline{\wedge}}} APC'D'$. Then $AP//C'D'$; that is, the joins $OC$ and $OD$ each meets one of the two segments $AP$, contrary to our definition of $A$. Hence in fact no three of the $n$ points can form a triangle, but all must be collinear.

### EXERCISES

1. Assuming $AB//CD$, name the four segments into which $A$, $B$, $C$, $D$ decompose the line.

2. Show that the relation $S(ABD) = S(BCD)$ implies $S(ABD) = S(ACD)$.

3. Given $n$ points, not all collinear, prove that by joining every two of them we obtain at least $n$ distinct lines. (P. Erdös.)

**3·4 Ordered Correspondence.** We have already described the concept of a correspondence between two ranges, illustrating it by the particular correspondence called perspectivity, which (by Axiom 3·16) preserves the relation of separation and consequently cyclic order and the distinction of sense. But this property of a perspectivity is shared by many other kinds of correspondence. Let us use the name *ordered correspondence* whenever the relation of separation is preserved. That is to say, the characteristic property of an ordered correspondence $X \to X'$ is that, if the relation $AB//CD$ holds for four positions $A$, $B$, $C$, $D$ of $X$, then the relation $A'B'//C'D'$ holds for the corresponding positions of $X'$. It follows that segments correspond to segments, and intervals to intervals.

Any point $M$ that coincides with its corresponding point $M'$ is called an *invariant* point. (Some authors prefer to call it a *double* point.) For instance, a perspectivity between two ranges has just one invariant point, where the two lines intersect. Of course, there are some ordered correspondences that have no invariant points. On the other hand, there may be more than one invariant point, but in such a case it is obvious that both ranges must be on the same line.

Any ordered correspondence preserves the distinction of sense, *i.e.*, the relation S(*ABC*) = S(*DEF*) implies S(*A'B'C'*) = S(*D'E'F'*); but in the case of a correspondence between superposed ranges (on one line) the question arises as to whether the sense *ABC* agrees or disagrees with the sense *A'B'C'*. We call such a correspondence *direct* or *opposite* according as

$$S(ABC) = S(A'B'C') \quad \text{or} \quad S(ABC) \neq S(A'B'C')$$

Whichever relation holds for one triad of points must still hold for any other, in view of the above remark about S(*ABC*) and S(*DEF*). In particular, the identity (namely, $X \to X$) is direct.

Fig. 3·4A

A particularly important kind of opposite correspondence is described in the following theorem:

**3·41** *The correspondence between the points of a range and their harmonic conjugates wo two fixed points M and N is an opposite correspondence with invariant points M and N.*

*Proof:* Since a perspectivity preserves order, so does the resultant or *product* of any sequence of perspectivities. To prove that the correspondence between harmonic conjugates wo *M* and *N* is ordered, we exhibit it as the product of three perspectivities with centers *Q*, *M*, *R*, in the notation of Fig. 3·4A. Here *Q* and *R* are any fixed points collinear with *M*. A variable point *X* on the line *MN* determines

$$P = NR \cdot QX, \qquad S = MP \cdot NQ, \qquad X' = MN \cdot RS$$

and we have

$$MNX \overset{Q}{\underset{\overline{\wedge}}{=}} RNP \overset{M}{\underset{\overline{\wedge}}{=}} QNS \overset{R}{\underset{\overline{\wedge}}{=}} MNX'$$

Thus the correspondence $X \to X'$ is ordered and has *M* and *N* as invariant points. Finally, it is opposite, since

$$\text{S}(MNX) \ne \text{S}(MNX')$$

by 3·21 and 3·32.

This enables us to prove the following:

**3·42**   *Two pairs of harmonic conjugates wo M and N cannot separate each other.*

*Proof:* Suppose that H($MN$, $AB$) and H($MN$, $CD$). Then $A$, $B$, $C$ are three positions of $X$ in the above correspondence, and the respective positions of $X'$ are $B$, $A$, $D$. Since the correspondence is opposite, we have $\text{S}(ABC) \ne \text{S}(BAD)$, that is, $\text{S}(ABC) = \text{S}(ABD)$. By 3·32 this means that $A$ and $B$ do not separate $C$ and $D$.

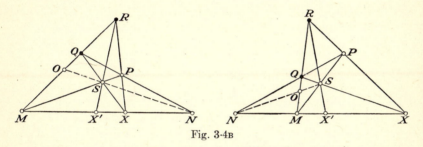

Fig. 3·4B

In proving 3·41 we exhibited the correspondence $X \to X'$, where H($MN$, $XX'$), as the product of three perspectivities (see Fig. 3·4A). By naming one further point $O = MQ \cdot SX$ we can reduce the number of perspectivities to two. For since H($MQ$, $OR$), $O$ is a fourth fixed point on the line $MQR$, and we have

$$MNX \overset{O}{\underset{\barwedge}{\phantom{x}}} QNS \overset{R}{\underset{\barwedge}{\phantom{x}}} MNX'$$

Interchanging $N$ and $X$, as in Fig. 3·4B, we obtain a new correspondence $X \to X'$, where now H($MX$, $NX'$):

**3·43**   *When* H($MX$, $NX'$), *where* $M$ *and* $N$ *are fixed,* $X \to X'$ *is a direct correspondence with invariant points* $M$ *and* $N$.

*Proof:* Using three fixed points $O$, $Q$, $R$ (outside the line $MN$) such that H($MQ$, $OR$), we observe that $QX$ meets $RX'$ in a point $S$ on $ON$, and

$$MNX \overset{Q}{\underset{\barwedge}{\phantom{x}}} ONS \overset{R}{\underset{\barwedge}{\phantom{x}}} MNX'$$

In this case $\text{S}(MNX) = \text{S}(MNX')$; for otherwise we should have $MN//XX'$, whereas the relation H($MX$, $NX'$) implies $MX//NX'$. Thus the correspondence is direct.

In ordinary analytic geometry a point on the $x$ axis is located by means of an abscissa $x$, which measures its distance to the right of the origin (so that points to the left of the origin have negative abscissas). A correspondence $X \to X'$ on this axis is represented by a correspondence of abscissas:

$$x \to x' = f(x)$$

where $f(x)$ is a single-valued function such that, for any given $x'$, the equation $f(x) = x'$ has a unique solution. Since we are considering the *projective* line, we must include $\infty\ (= -\infty)$ as a possible value for $x$ or $x'$. If $x$ and $x'$ become infinite together, the point at infinity is invariant; if not, there will be a finite $x' = f(\infty)$ and a finite $x$ for which $f(x) = \infty$. In the special case when $f(x)$ is $x$ itself, the correspondence is the identity. If $M$ and $N$ have abscissas 0 and $\infty$, the correspondences considered in 3·41 and 3·43 are, respectively, $x' = -x$ and $x' = \frac{1}{2}x$. Our axioms have been chosen so as to enable us to develop the same theory without having recourse to analysis.

### EXERCISES

1. Show that the correspondence $x \to x'$ is ordered if $x'$ is a differentiable function whose derivative $dx'/dx$ never changes sign. It is direct if $dx'/dx > 0$ almost everywhere (*i.e.*, except where $x'$ is infinite, and possibly at some isolated places where the derivative may vanish) and opposite if $dx'/dx < 0$ almost everywhere. *Hint:* Compare the signs of $(a' - b')/(a - b)$ and $(b' - c')/(b - c)$, using the mean-value theorem.

2. Assuming the function $x' = f(x)$ to be continuous (except where it becomes infinite), show that the graph $y = f(x)$ is either all in one piece with no asymptote or in two pieces with two asymptotes, one horizontal and one vertical. Which points on the graph represent invariant points of the correspondence?

3. Show that $x \to x^3$ is a direct correspondence with four invariant points (where $x = -1, 0, 1, \infty$).

**3·5  Continuity.** To get a picture of what is happening in an ordered correspondence $X \to X'$ on one line, think of a circular race track that two runners agree to run all round, starting at the same time and finishing at the same time, never stopping or turning back but otherwise free to go as fast or slow as they please. Then $X$ and $X'$ are the respective positions of the two runners at any instant. The correspondence is direct or opposite according as the runners are going in the same direction or in opposite directions. An invariant point occurs where the runners meet or where one overtakes the other. In the direct case this may happen any number of times, even infinitely often, for the runners might remain side by side for awhile (or even for the whole journey, when the correspondence is the identity). Thus there may be any number of invariant points, from none at all to infinitely many. But in the opposite case a little thought reveals that the runners will meet exactly twice before each returns to his own start-ing point (or if they started from the same point, they will meet · once more elsewhere). This means that every opposite correspond-ence should have exactly two invariant points; but we cannot prove this rigorously without introducing one further assumption, such as the following:

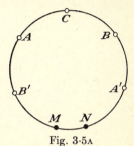

Fig. 3·5A

**3·51** Axiom of Continuity. *If an ordered correspondence relates an interval* $\overline{AB}/C$ *to an interior interval* $\overline{A'B'}/C$, *then the latter contains an invariant point M such that there is no invariant point between A and M (in $\overline{AB}/C$).*

This is obvious when we think of the race-track: $X$ runs from $A$ to $B$ while $X'$ runs *over part of the same ground* from $A'$ to $B'$; $M$ is the first point where they meet.

If the correspondence is opposite, the last clause of the axiom (after "point $M$") is superfluous: $M$ is the *only* invariant point in $\overline{AB}/C$. For, two invariant points (such as the $M$ and $N$ of Fig. 3·5A) would determine a segment whose sense is preserved.

Fig. 3·6A

**3·6　Invariant Points.**　We are now ready to prove the following:

**3·61**　*Every opposite correspondence has exactly two invariant points.*

*Proof:* Since the identity is direct, any opposite correspondence admits a point $A$ that is not invariant. Suppose the correspondence relates $A$ to $A'$ and $A'$ to $A''$. Choose a point $C$ such that $AA'//A''C$ (or, if $A''$ coincides with $A$, take *any* new point $C$). Then the given opposite correspondence relates $\overline{AA'}/C$ to the interior interval $\overline{A'A''}/C$,

---

\* We use the interval, rather than the segment $AB/C$, to cover the possibility of $B'$ coinciding with $B$, in which case $M$ might also coincide with $B$.

as in Fig. 3·6A. Hence there is just one invariant point $M$ in $AA'/C$. Similarly, there is a second invariant point $N$ in the supplementary segment $AA'/M$; for the inverse correspondence (namely $X' \rightarrow X$) relates $A'A''/M$ to the interior interval $\overline{AA'}/M$.

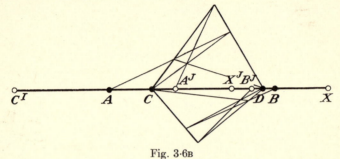

Fig. 3·6B

We saw, in 3·42, that the relations $H(MN, AB)$ and $H(MN, CD)$ preclude $AB//CD$. This theorem has an important converse:

**3·62** *If $AB$ and $CD$ are two pairs of points that are collinear but do not separate each other, then there exist points $M$ and $N$ such that*

$$H(AB, MN) \quad and \quad H(CD, MN).$$

*Proof:* Any point $X$ has a harmonic conjugate $X^I$ wo $A$ and $B$ and a harmonic conjugate $X^J$ wo $C$ and $D$: in symbols,

$$H(AB, XX^I) \quad and \quad H(CD, XX^J).$$

While $X$ runs from $A$ to $B$ over the interval $\overline{AB}/C$, $X^I$ runs from $A$ to $B$ over the supplementary interval $\overline{AB}/C^I$, which includes $D$ as well

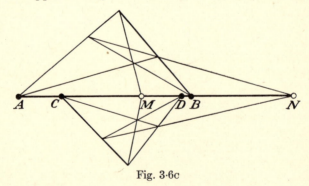

Fig. 3·6c

as $C$, since the pairs $AB$ and $CD$ do not separate each other (see Fig. 3·6B). Meanwhile, $X^J$ runs from $A^J$ to $B^J$ over part of the same interval. Now consider the combined correspondence $X^I \rightarrow X^J$. This relates the interval $\overline{AB}/C^I$ to the interior interval $\overline{A^JB^J}/C$. By Axiom 3·51 the latter interval contains an invariant point $M$, which can equally well be called $N^I$ or $N^J$, since it is the harmonic conjugate of some point $N$ wo either of the pairs $AB$, $CD$ (see Fig. 3·6c).

The theorem we have just proved is especially significant since it would enable us to define separation in terms of incidence, instead of taking separation to be a second undefined relation. In fact, we could define $AB//CD$ to mean that there is no common pair of harmonic conjugates wo the two pairs $AB$ and $CD$. Then we could prove 3·12, 3·13, and 3·16 as theorems, leaving only three axioms of order. This idea is due to Pieri, whose exposition was praised by Russell[*] in the following words: "This is, in my opinion, the best work on the present subject." Actually, instead of $AB/C$, Pieri defined the supplementary segment $(A\check{C}B)$, containing $C$; and he took 3·22 as an axiom. His definition[†] may be expressed thus:

$(ACB)$ *is the locus of the harmonic conjugate of $C$ wo a variable pair of distinct points that are harmonic conjugates wo $A$ and $B$.*

Adding to the segment its end points $A$ and $B$, we obtain an interval, say $(\overline{ACB})$; and the relation $AB//CD$ means that $D$ does not belong to this interval. Thus Pieri reduced the undefined relations to incidence alone and reduced the axioms of order to the following three:

(1) *If $D$, on $AB$, does not belong to $(\overline{ACB})$, it belongs to $(ABC)$.*

(2) *If $D$ belongs to both $(ABC)$ and $(BAC)$, it cannot belong to $(ACB)$.*

(3) *If $D$ belongs to $(ACB)$ and $E$ to $(ADB)$, then $E$ belongs to $(ACB)$.*

On the other hand, this simplification is to some extent illusory, as these axioms would be quite complicated if we expressed them directly in terms of incidence. Now, which is preferable: a number of simple axioms involving two undefined relations, or fewer but far more complicated axioms involving only one such relation? The answer is a matter of taste.

### EXERCISES

1. Prove that if $D$ belongs to $(ACB)$, $(ACB) = (ADB)$.

*Hint:* By Pieri's definition of a segment, $C$ belongs to $(ADB)$. By (3), every point of $(ADB)$ belongs to $(ACB)$. By the same axiom with $C$ and $D$ interchanged, every point of $(ACB)$ belongs to $(ADB)$.

2. Deduce 3·11, 3·14, and 3·15 from Pieri's axioms.

*Hint:* If 3·11 were not true, $(\overline{ACB})$ would cover the whole line. Similarly, so would $(\overline{ABC})$ and $(\overline{BAC})$. Any fourth point $D$ would belong to all three of the segments $(ACB)$, $(ABC)$, $(BAC)$, contradicting (2). As for 3·14, this is a simple restatement of (2). To prove 3·15 we may argue as follows:[‡] Since

[*] Ref. 36, p. 382.

[†] Ref. 28, p. 24.

[‡] This solution is due to R. G. E. Epple, a graduate student at the University of Southern California.

$AB//CD$ and $AC//BE$, $D$ does not belong to $(\overline{AC}\overline{B})$, nor $E$ to $(\overline{AB}\overline{C})$. Hence, by (1), $E$ belongs to $(ACB)$ while $D$ does not. Interchanging $D$ and $E$ in (3), we conclude that $D$ cannot belong to $(AEB)$, nor even to $(\overline{AE}\overline{B})$. Hence $AB//DE$.

**3·7   Order in a Pencil.**   If $a$, $b$, $c$, $d$ are four concurrent lines meeting another line in points $A$, $B$, $C$, $D$ such that $AB//CD$, then we say

$$ab//cd$$

By Axiom 3·16 this definition for separation of line pairs is independent of the chosen section $ABCD$. We can easily dualize all the results of the present chapter; *e.g.*, the dual of a segment is an angle. If $a$, $b$, $c$ are three lines through a point $O$, we can distinguish the two senses of rotation about $O$ as S($abc$) and S($cba$). We can define an *ordered correspondence* between two pencils. If the pencils have the same center,

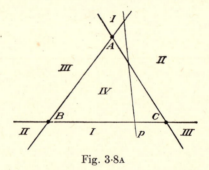

Fig. 3·8A

the correspondence may be *direct* or *opposite*. By 3·61 every opposite correspondence between pencils has exactly two invariant lines.

**3·8   The Four Regions Determined by a Triangle.**   We saw, in 3·31, that two points decompose their line into just two segments. Dually, two lines decompose the pencil to which they belong into two angles; or we may say that they decompose the whole plane into two angular *regions*. ("Vertically opposite" angles belong to the same region, since the line at infinity forms no barrier between them.) A third line, not concurrent with the first two, penetrates both regions. Hence:

**3·81**   *Three lines that form a triangle decompose the whole plane into four regions.*\*

A fourth line, not through a vertex of the triangle, is decomposed by the three lines into three segments, one in each of some three of the

---

\* For a more extended account (though without the notation $ABC/p$) see Veblen and Young (Ref. 43, p. 53).

four regions. Hence there is just one of the four regions that the new line fails to penetrate. This suggests a notation for distinguishing the four regions formed by lines $BC$, $CA$, $AB$: the region *not* penetrated by a line $p$ is denoted by

$$ABC/p$$

(In Fig. 3·8A, this is region *III*.)

### EXERCISES

1. Observe that the interior of the ordinary triangle $ABC$ of affine geometry may be described as $ABC/o$, where $o$ is the line at infinity.

2. In Euclidean geometry, three lines forming a triangle are tangents to four circles (inscribed and escribed), one in each of the four regions.

3. In ordinary analytic geometry the coordinate axes $OX$, $OY$ and the line at infinity may be regarded as forming a triangle that decomposes the plane into the four *quadrants*. Show that each of these may be denoted by $OXY/p$, where $p$ is one of the four lines

$$\pm x \pm y + 1 = 0$$

4. Show that four lines of general position (forming a complete quadrilateral) decompose the plane into seven regions: four triangular (bounded by three segments) and three quadrangular (bounded by four segments).

# CHAPTER 4

# *ONE-DIMENSIONAL PROJECTIVITIES*

The present chapter is concerned with the most important kind of ordered correspondence: the projectivity, which may be defined either as the product of several perspectivities or as a correspondence that preserves harmonic sets. The first definition, due to Poncelet, has been adopted by Veblen, Baker, and other authors; it has the advantage of remaining valid in complex geometry. But this book follows Enriques in using the second definition, due to von Staudt, which generalizes more readily to two (or more) dimensions. It is an immediate consequence of 2·82 that every Poncelet projectivity is a von Staudt projectivity, and we shall prove in Sec. 4·2 that every von Staudt projectivity (in real geometry) is a Poncelet projectivity. Thus from that point on the two treatments coincide.

**4·1 Projectivity.** The notion of correspondence extends easily from the line to the plane. By a two-dimensional correspondence $X \to X'$ we mean a rule for associating every point $X$ with every point $X'$ so that there is exactly one $X'$ for each $X$ and exactly one $X$ for each $X'$. A correspondence between lines, $x \to x'$, is defined similarly.*

A *collineation* is the special case where collinear points correspond to collinear points, and consequently concurrent lines to concurrent lines; *i.e.*, ranges correspond to ranges, and pencils to pencils. Thus a collineation preserves incidences: point $X'$ lies on line $x'$ if and only if point $X$ lies on line $x$. The range of points $X$ on a given line $x$ corresponds to a range of points $X'$ on the corresponding line $x'$. Four positions of $X$ forming a harmonic set correspond to four positions of $X'$ forming a harmonic set; for any quadrangle used in constructing the first set corresponds to a quadrangle having the same relation to the second set. This suggests the following one-dimensional analogue:

---

* If we had not restricted our geometry to two dimensions (by means of Axiom 2·24), we could just as easily have defined a correspondence between the points (or lines) of two distinct planes.

A *projectivity* between two ranges is a correspondence that preserves the harmonic relation. In other words, if the relation H($AB$, $CD$) holds for four positions $A$, $B$, $C$, $D$ of $X$, then the relation H($A'B'$, $C'D'$) holds for the corresponding positions of $X'$. The established notation, invented by von Staudt,* is

$$X \barwedge X'$$

Thus the relations $ABCD \barwedge A'B'C'D'$ and H($AB$, $CD$) imply H($A'B'$, $C'D'$).

By 2·82 the perspectivity $X \doublebarwedge X'$ is a special case of the projectivity $X \barwedge X'$. We may now write 2·71 in the concise form

$$ABCD \barwedge BADC \barwedge CDAB \barwedge DCBA$$

(for any four collinear points).

We also define a projectivity $x \barwedge x'$ between the lines of two pencils: if H($ab$, $cd$) holds for four positions $a$, $b$, $c$, $d$ of $x$ in the first pencil, then H($a'b'$, $c'd'$) holds for the corresponding four positions of $x'$ in the second.

The following theorem will enable us to apply to projectivities some of the results already obtained for ordered correspondences (*e.g.*, 3·61):

**4·11**　*Every projectivity is an ordered correpondence.* In other words, if $ABCD \barwedge A'B'C'D'$ and $AB//CD$, then $A'B'//C'D'$.

*Proof:* Suppose, if possible, that $ABCD \barwedge A'B'C'D'$ and $AB//CD$ but not $A'B'//C'D'$. Then by 3·62 there exist points $M'$ and $N'$ such that H($A'B'$, $M'N'$) and H($C'D'$, $M'N'$). These two points of the second range correspond to points $M$ and $N$ of the first, such that H($AB$, $MN$) and H($CD$, $MN$). By 2·83 this means that H($MN$, $AB$) and H($MN$, $CD$). But we have assumed $AB//CD$; thus 3·42 is contradicted.

The next theorem shows a radical departure from the general ordered correspondence (which, if direct, may have any number of invariant points):

**4·12**　*A projectivity having more than two invariant points can only be the identity.*†

*Proof:* We shall obtain a contradiction by supposing that a given projectivity has three invariant points $A$, $B$, $C$ and a noninvariant point $P$, so that $ABCP \barwedge ABCP'$ with $P \neq P'$. Let the points $A$, $B$, $C$ be named in such an order that $P$ lies in the segment $AB/C$ and $P'$ in $PB/C$ (see Fig. 4·1A). The projectivity relates the interval $\overline{PB}/C$ to

* Ref. 40, p. 49.
† von Staudt (Ref. 40, p. 50, §106).

the interior interval $\overline{P'B}/C$. Hence by Axiom 3·51 the latter interval contains a "first" invariant point $M$ (admitting no invariant points between $P$ and $M$). Similarly the inverse projectivity $(X' \underset{\wedge}{\phantom{|}} X)$ relates $\overline{AP'}/C$ to $\overline{AP}/C$, which consequently contains a "last" invariant point $N$ (admitting no invariant points between $N$ and $P'$). Since the segments $NP'/C$ and $PM/C$ overlap, we can assert that the segment $NM/C$ is entirely free from invariant points.

Let $D$ be the harmonic conjugate of $C$ wo $M$ and $N$, and suppose $D \underset{\wedge}{\phantom{|}} D'$. Since $MNCD \underset{\wedge}{\phantom{|}} MNCD'$, the relation H($MN$, $CD$) implies H($MN$, $CD'$). Hence, by 2·51, $D = D'$, and $D$ is an invariant point in the forbidden segment $MN/C$, which is absurd. Thus there cannot really be three invariant points (unless *every* point is invariant).

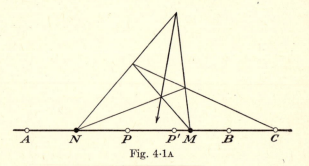

Fig. 4·1A

**4·2  The Fundamental Theorem of Projective Geometry.** The following theorem derives it's name from the fact that it opens the way to the most characteristic developments of our subject. Its strength will be seen in the ease with which the remaining theorems of this chapter can be proved. Moreover, it enables us to construct any given projectivity as a product of perspectivities, thus reconciling the treatments of Poncelet and von Staudt.

**4·21**  The Fundamental Theorem: *A projectivity is determined when three points of one range and the corresponding three points of the other are given.*[*]

*Proof:* Suppose we are given three points $A$, $B$, $C$ of one range and corresponding points $A'$, $B'$, $C'$ of the other. We wish to construct a projectivity $X \underset{\wedge}{\phantom{|}} X'$ such that $ABC \underset{\wedge}{\phantom{|}} A'B'C'$ and to establish its uniqueness.

If the two ranges are on distinct lines, as in Fig. 4·2A, one simple construction is obtained by letting $R$, $S$, $C_0$ denote the points where

---

[*] von Staudt (Ref. 40, p. 52, §110).

the respective lines $AA'$, $BB'$, $BA'$ meet $CC'$. Then any point $X$ on $AB$ determines $X'$ on $A'B'$ by means of the two perspectivities

$$ABCX \overset{R}{\barwedge} A'BC_0X_0 \overset{S}{\barwedge} A'B'C'X'$$

If the two ranges are on one line, we use a quite arbitrary perspectivity $ABC \barwedge A_1B_1C_1$ to obtain a range on another line and then relate $A_1B_1C_1$ to $A'B'C'$ by the above construction.

Fig. 4·2A                    Fig. 4·2B

To establish the uniqueness of this projectivity, we have to prove that a different construction (*e.g.*, by joining $AB'$ instead of $BA'$) would yield the same $X'$ for a given $X$. Suppose one construction gives

$$ABCX \overline{\wedge} A'B'C'X'$$

while another gives

$$ABCX \overline{\wedge} A'B'C'X'_1$$

Then by combining the two constructions we obtain

$$A'B'C'X' \overline{\wedge} A'B'C'X'_1$$

This combined projectivity has three invariant points $A'$, $B'$, $C'$; hence, by 4·12, $X'_1$ must coincide with $X'$.

**4·22  Corollary:** *Any projectivity can be constructed as a product of perspectivities, the number of which can be reduced to three. If the two ranges are on distinct lines, two perspectivities suffice.*\*

In one important case a single perspectivity suffices:

**4·23**  *If a projectivity between ranges on two distinct lines has an invariant point, it is merely a perspectivity.*†

---

\* For the direct deduction of this theorem from Poncelet's definition of a projectivity, see Robinson (Ref. 32, pp. 28–31) or Hodge and Pedoe (Ref. 20, pp. 218–224).

† von Staudt (Ref. 40, p. 51, §108).

*Proof:* Of course, the invariant point $O$ belongs to both ranges; thus it must be the point of intersection of the two lines, as in Fig. 4·2B. Let $B$ and $C$ be any other points of the first range, $B'$ and $C'$ the corresponding points of the second. Then we have $OBC \mathbin{\overline{\wedge}} OB'C'$. But

$$OBC \mathbin{\overset{S}{\overline{\wedge}}} OB'C' \quad \text{where} \quad S = BB' \cdot CC'$$

By the fundamental theorem this perspectivity is the same as the given projectivity: the join of two corresponding points always passes through this same point $S$.

### EXERCISE

If the sides of a variable triangle pass through three fixed collinear points, while two vertices run along fixed lines, prove that the third vertex will run along a third fixed line concurrent with the other two. (This is Pappus's porism, which was the inspiration for much of Maclaurin's work on loci, beginning in 1722.)

*Hint:* Either use the dual of 4·23 or apply 2·26 to two positions of the variable triangle.

Fig. 4·3A          Fig. 4·3B

**4·3  Pappus's Theorem.** The theorem we are about to prove is especially significant because in some treatments it is taken as an axiom, instead of 3·51. The resulting geometry is more general, as it can be developed without any appeal to continuity.

**4·31**  Pappus's Theorem: *If alternate vertices of a hexagon lie on two lines, the three pairs of opposite sides meet in three collinear points.*

*Proof:*\* Let $AB'CA'BC'$ be the hexagon, so that the points to be proved collinear are

$$L = BC' \cdot CB', \quad M = CA' \cdot AC', \quad N = AB' \cdot BA'$$

as in Fig. 4·3A. Using further points

$$J = AC' \cdot BA', \quad K = BC' \cdot CA', \quad O = AB \cdot A'B'$$

---

\* O'Hara and Ward (Ref. 26, p. 53).

we have

$$A'NJB \overset{A}{\underset{\barwedge}{=}} A'B'C'O \overset{C}{\underset{\barwedge}{=}} KLC'B$$

Thus $B$ is an invariant point of the projectivity $A'NJ \underset{\barwedge}{\to} KLC'$. By 4·23 this is a perspectivity, namely, $A'NJ \overset{M}{\underset{\barwedge}{=}} KLC'$ (since the joins $A'K$ and $JC'$ pass through $M$). Hence $NL$ passes through $M$.

Pappus's theorem suggests a more symmetrical construction (Fig. 4·3c) to replace Fig. 4·2A. Given four points $A$, $B$, $C$, $X$ on one line and three points $A'$, $B'$, $C'$ on another, we can locate $X'$ such that

Fig. 4·3c

$ABCX \underset{\barwedge}{\to} A'B'C'X'$ as the point $A'B' \cdot AF$, where $F$ is $XA' \cdot o$, and $o$ is the "Pappus line"

$$(CA' \cdot AC')(AB' \cdot BA')$$

sometimes called the *axis* of the projectivity. To see this, let $G$ be the point where $AA'$ meets the axis $o$. Then

$$ABCX \overset{A'}{\underset{\barwedge}{=}} GNMF \overset{A}{\underset{\barwedge}{=}} A'B'C'X'$$

It is remarkable that Pappus's theorem, when used as an axiom, can take the place of Axiom 2·25. In fact, *Desargues' theorem can be deduced from 2·21 to 2·24 and 4·31.* The following is a simplified version of Hessenberg's proof:*

Using the notation of Fig. 1·5B or the frontispiece, name the four extra points

$$S = PR \cdot Q'R', \qquad T = PQ' \cdot RR', \qquad U = PQ \cdot OS, \qquad V = P'Q' \cdot OS$$

* Pasch and Dehn (Ref. 27, p. 227); Baker (Ref. 3, pp. 25–26); Hodge and Pedoe (Ref. 20, p. 272).

Then the Pappus hexagon $QROSQ'P$ makes $A$, $T$, $U$ collinear,

$P'R'OSPQ'$ makes $B$, $T$, $V$ collinear,

and     $Q'SPUTV$ makes $A$, $B$, $C$ collinear.

## EXERCISES

1. Show that Pappus's theorem is its own dual (see Fig. 4·3B).

2. Let the points $A$, $B$, $C$, $A'$, $B'$, $C'$, $L$, $M$, $N$ of Fig. 4·3A be renamed $A_1$, $B_1$, $C_1$, $A_2$, $B_2$, $C_2$, $A_3$, $B_3$, $C_3$. Observe that $A_i$, $B_j$, $C_k$ are collinear whenever $i + j + k$ is a multiple of 3.*

3. Given a triangle $A_1A_2A_3$ and two points $B_1$, $B_2$, locate a point $B_3$ such that the lines $A_1B_1$, $A_2B_3$, $A_3B_2$ are concurrent, while also $A_1B_3$, $A_2B_2$, $A_3B_1$ are concurrent. Prove that then the lines $A_1B_2$, $A_2B_1$, $A_3B_3$ are concurrent. (In other words, if two triangles are doubly perspective, they are triply perspective.)†

4. Show that the Pappus configuration of nine points and nine lines may be regarded (in six ways) as consisting of a cycle of three triangles, such that the three sides of each pass through the three vertices of the next.‡ *Hint:* Let one of the triangles be $ABN$ (or $A_1B_1C_3$).

5. Let the axis of $ABC \overline{\underline{\wedge}} A'B'C'$ (Fig. 4·3c) meet $AB$ in $Q$ and $A'B'$ in $O'$. Show that $ABCOQ \overline{\underline{\wedge}} A'B'C'O'O$.

**4·4  Classification of Projectivities.** A projectivity on one line may be either *direct* (sense-preserving) or *opposite* (sense-reversing). The identity is, of course, direct; by 4·12, no other projectivity can have more than two invariant points.

A projectivity having no invariant point is said to be *elliptic.*

A projectivity having one invariant point is said to be *parabolic.*

A projectivity having two invariant points is said to be *hyperbolic.*

(These names will be justified when we come to consider affine geometry, where the various kinds of conic have 0, 1, or 2 points at infinity.)

By 3·61, every opposite projectivity is hyperbolic; therefore every elliptic or parabolic projectivity is direct. Thus the two methods of classification are related as in the following:

*Table of Projectivities on One Line*

| Direct | | | Opposite |
|---|---|---|---|
| The identity ($\infty$) | Elliptic (0) | Parabolic (1) | Hyperbolic (2) |

(The numbers of invariant points are given in parentheses.)

* Levi (Ref. 24, p. 108).

† Veblen and Young (Ref. 42, p. 100).

‡ Hessenberg (Ref. 18, p. 69).

This shows that, apart from the identity, there are four possible kinds of projectivity (for superposed ranges): elliptic, parabolic, direct hyperbolic, and opposite. We proceed to prove that all four kinds actually exist.

Special hyperbolic projectivities of the two kinds have already appeared in 3·43 and 3·41.

An instance of an elliptic projectivity is afforded by

$$ABC \barwedge BCA$$

where $A$, $B$, $C$ are any three collinear points. These points themselves are obviously not invariant, and each of the three segments $BC/A$, $CA/B$, $AB/C$ is related to another one; hence there is no place for an

Fig. 4·4A

invariant point anywhere. The actual construction for an elliptic projectivity requires the full allowance of *three* perspectivities. For if a projectivity $X \barwedge X'$ on one line is the product of two perspectivities

$$X \barwedge X_0 \barwedge X'$$

there must be an invariant point where the line of $X$'s meets the line of $X_0$'s. Conversely,

**4·41**   *Every parabolic or hyperbolic projectivity (with a given invariant point) can be constructed as the product of two perspectivities.*

*Proof:* By the fundamental theorem 4·21, a projectivity having an invariant point $M$ is uniquely determined by the relation

$$MAB \barwedge MA'B'$$

Choose any two points $A_0$ and $B_0$ collinear with $M$, and construct

$$R = AA_0 \cdot BB_0, \qquad S = A_0A' \cdot B_0B'$$

as in Fig. 4·4A. Then we can locate the $X'$ for a given $X$ by means of the two perspectivities

$$MABX \underset{\overline{\wedge}}{\overset{R}{}} MA_0B_0X_0 \underset{\overline{\wedge}}{\overset{S}{}} MA'B'X'$$

Here $M$ is the given invariant point. Any other invariant point $N$ must project from $R$ and $S$ into the same point $N_0$ on $A_0B_0$; hence it must lie on $RS$. Thus the projectivity is parabolic if $RS$ passes through $M$, and hyperbolic otherwise.

By 4·21, a hyperbolic projectivity is determined when both invariant points and one pair of corresponding points are given:

$$MNA \overline{\wedge} MNA'$$

Such a projectivity exists for any four collinear points $M$, $N$, $A$, $A'$. To construct it, choose any two points $R$ and $S$ collinear with $N$,

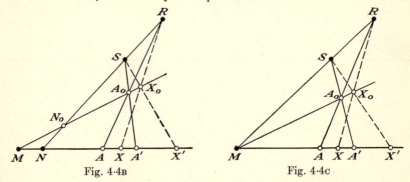

Fig. 4·4B                    Fig. 4·4C

locate $A_0 = AR \cdot A'S$, and use the line $MA_0$ as before. For any $X$ on $MA$ we have

$$MNAX \underset{\overline{\wedge}}{\overset{R}{}} MN_0A_0X_0 \underset{\wedge}{\overset{S}{}} MNA'X'$$

as in Fig. 4·4B.

The pair $AA'$ may or may not separate the pair $MN$. In the latter case, by 3·32, S($MNA$) = S($MNA'$). Hence:

**4·42** *The hyperbolic projectivity $MNA \overline{\wedge} MNA'$ is opposite if $MN//AA'$ and direct otherwise.* *

The above construction remains valid when $N$ coincides with $M$ (so that the projectivity is no longer hyperbolic but parabolic, as in Fig. 4·4C). In this case, if $X$ is taken at $A'$, the quadrangle $RSA_0X_0$ gives H($MA'$, $AX'$). Conversely, such a figure can be reconstructed from any harmonic set $MA'$, $AA''$, and then $M$ is the only invariant point of the projectivity $MAA' \overline{\wedge} MA'A''$. Hence:

* Enriques (Ref. 11, p. 101).

**4·43**  *The projectivity $MAA' \barwedge MA'A''$ is parabolic if* $H(MA', AA'')$ *and hyperbolic otherwise.*

**4·44**  Corollary: *A parabolic projectivity is determined when its invariant point and one pair of corresponding points are given.*

Such a projectivity is naturally denoted by

$$MMA \barwedge MMA'$$

therefore we may say that the relation $H(MA', AA'')$ is equivalent to $MMAA' \barwedge MMA'A''$.

This notation is justified by its transitivity:

**4·45**  *The product of two parabolic projectivities having the same invariant point is another such parabolic projectivity (if it is not merely the identity).*

*Proof:* Clearly, the common invariant point of the two given projectivities is still invariant for the product. If any other point $A$ were invariant, too, the first projectivity would take $A$ to some different point $A'$ and the second would take $A'$ back to $A$. By 4·44 the second would then be just the inverse of the first. Hence, apart from that trivial case, the product is parabolic with the same invariant point.

### EXERCISES

1. Which part of Fig. 4·4A is *direct*?
2. Draw a figure to illustrate $MMAA'A'' \barwedge MMA'A''A'''$.

**4·5  Periodic Projectivities.**  Suppose a given projectivity relates $X$ to $X'$, $X'$ to $X''$, and so on:

$$XX'X'' \cdots X^{(n-1)} \barwedge X'X''X''' \cdots X^{(n)}$$

If $X^{(n)}$ coincides with $X$ for three (and therefore all) positions of $X$, the projectivity is said to be *periodic* and the smallest $n$ for which this happens is called the *period*. Thus the identity is of period 1, the correspondence between harmonic conjugates wo $M$ and $N$ (see 3·41) is of period 2, and the elliptic projectivity $ABC \barwedge BCA$ (Sec. 4·4) is of period 3.

### EXERCISES

1. Show that, if $H(AC, BD)$, $ABC \barwedge BCD$ is an elliptic projectivity of period 4.

2. Show that a parabolic projectivity cannot be periodic. *Hint:*

$$S(MXX') = S(MX'X'') = S(MX''X''') = \cdots$$

3. Prove that every periodic hyperbolic projectivity is opposite.

4. Prove that the only possible period for a hyperbolic projectivity is 2. (Otherwise the "squared" projectivity $X \barwedge X''$ would be periodic, hyperbolic, and direct.)

**4·6  Involution.**  Desargues defined an involution as the relation between pairs of points on a line whose distances from a fixed point have a constant product (positive or negative). The following projective definition is due to von Staudt.*

An *involution* is a projectivity of period 2:

$$XX' \barwedge X'X$$

It is remarkable that this relation holds for all positions of $X$ if it holds for any one position; in other words,

**4·61**  *A projectivity that interchanges two points is necessarily an involution.*

*Proof:*  Suppose we are given $AA' \barwedge A'A$. Consider any point $X$, and suppose $X \barwedge X'$. By the fundamental theorem 4·21, the given projectivity is the only one in which

$$AA'X \barwedge A'AX'$$

By 2·71 there is a projectivity in which $AA'XX' \barwedge A'AX'X$. Hence this is the same as the given projectivity, and $XX'$ is a doubly corresponding pair. Since $X$ is quite arbitrary, this proves that the projectivity is an involution.

**4·62**  Corollary: *An involution is determined by any two of its pairs.*

*Notation:*  The involution $AA'BB' \barwedge A'AB'B$ is denoted by

$$(AA')(BB').$$

Either pair may be replaced by an invariant point repeated: the involution $AA'M \barwedge A'AM$ is denoted by $(AA')(MM)$.

**4·63**  *If an involution has one invariant point, it has another, and the involution is just the correspondence between harmonic conjugates wo these two points.*

*Proof:*  Consider the involution $(AA')(MM)$, and let $N$ be the harmonic conjugate of $M$ wo $A$ and $A'$. Then $N$ is also the harmonic conjugate of $M$ wo $A'$ and $A$. But the involution, being a projectivity, preserves the harmonic relation. Hence $N$ is a second invariant point (distinct from $M$, by 3·22). If another pair $XX'$ is used instead of $AA'$, we still obtain the same harmonic conjugate $N$, since otherwise the involution would have three invariant points.

* Ref. 40, pp. 119–120.

Corollary: *There is no parabolic involution.* *

Since the hyperbolic involution $AA'MN \barwedge A'AMN$ is completely determined by its two invariant points, we may denote it by

$$(MM)(NN).$$

From 3·41 we immediately deduce the following:
**4·64**    *Every hyperbolic involution is opposite.*

Thus every direct involution is elliptic, and for involutions the table on page 42 becomes simply

|     Direct     |     Opposite     |
| :------------: | :--------------: |
| Elliptic (0)   | Hyperbolic (2)   |

Any four collinear points $A$, $A'$, $B$, $B'$ determine an involution $(AA')(BB')$ or $AA'B \barwedge A'AB'$. If the involution is hyperbolic, the two pairs cannot separate each other (see 3·42). But if the involution is elliptic, it is direct,

$$\mathrm{S}(AA'B) = \mathrm{S}(A'AB') \neq \mathrm{S}(AA'B')$$

and $AA'//BB'$. Hence:

**4·65**    *The involution $(AA')(BB')$ is elliptic if $AA'//BB'$ and hyperbolic otherwise.*

We now recognize the points $M$ and $N$ of 3·62 as the invariant points of the hyperbolic involution $(AB)(CD)$.

The following criterion is often useful:

**4·66**    *A necessary and sufficient condition for three pairs $AA'$, $BB'$, $CC'$ to belong to an involution is $ABCC' \barwedge B'A'CC'$.*

*Proof:* If $CC'$ is a pair of $(AA')(BB')$, we have, by 2·71,

$$ABCC' \barwedge A'B'C'C \barwedge B'A'CC'$$

Conversely, the relation $ABCC' \barwedge B'A'CC' \barwedge A'B'C'C$ implies that the three pairs belong to an involution. (We may have $A = A'$ or $B = B'$, but the nature of the proof requires $C \neq C'$.)

---

* For some purposes it is convenient to admit the "degenerate involution" that relates every point $X$ to one fixed point $M$. The appropriate symbol is $(AM)(BM)$, where $A \neq B$.

Changing the notation, we may say that a necessary and sufficient condition for the pair $MN$ to belong to the involution $(AB')(BA')$ is

$$MNAB \,\overline{\wedge}\, MNA'B'$$

Of course, $(AB')(BA')$ is the same as $(AB')(A'B)$; hence:

**4·67**    *The relation* $MNAB \,\overline{\wedge}\, MNA'B'$ *is equivalent\* to*

$$MNAA' \,\overline{\wedge}\, MNBB'$$

If two involutions $(AA_1)(BB_1)$ and $(A'A_1)(B'B_1)$ have a common pair $MN$, the above remarks show that

$$MNAB \,\overline{\wedge}\, MNB_1A_1 \,\overline{\wedge}\, MNA'B'$$

Hence:

**4·68**    *If* $MN$ *is a pair of each of the involutions* $(AA_1)(BB_1)$ *and* $(A'A_1)(B'B_1)$, *it is also a pair of* $(AB')(BA')$.

One reason for the importance of involutions is apparent in the following theorem:

**4·69**    *Any one-dimensional projectivity may be expressed as the product of two involutions.*†

*Proof:* Let the given projectivity transform any noninvariant point $A$ into $A'$, and $A'$ into $A''$. Then its product with the involution $(AA'')(A'A')$ transforms the pair $AA'$ into $A'A$. Hence the product is itself an involution, and the given projectivity is the product of these two involutions, since the "square" of an involution is the identity. (In symbols, if the given projectivity is $T$, the first involution $I$, and the second $J$, we have $J = TI$, whence $JI = TI^2 = T$.)

### EXERCISES

1. Show that the relation $MNAB \,\overline{\wedge}\, MNBA$ is equivalent to H($AB, MN$).‡

2. If a hyperbolic projectivity has a pair of corresponding points that are harmonic conjugates wo the two invariant points, show that it must be an involution.

3. Given H($AA', MN$) and H($BB', MN$), prove that $A'B'$ is a pair of the involution $(AB)(MN)$.

4. Prove that two involutions, one or both elliptic, on the same line, always have a common pair of corresponding points. *Hint:* Consider the product $X \,\overline{\wedge}\, X^{IJ}$ of the two involutions $(XX^I)$ and $(XX^J)$. If one involution is elliptic and the other hyperbolic, the product is hyperbolic by 3·61. Let $M$ be one of the invariant points of this projectivity; then $M = M^{IJ}$ and $M^J = M^I$. On

---

\* von Staudt (Ref. 40, p. 59, §120). In this theorem it is the two *relations* that are equivalent (each implying the other); the two *projectivities* are, of course, distinct.

† Veblen and Young (Ref. 42, p. 224).

‡ von Staudt (Ref. 40, p. 58 §118).

the other hand, if both involutions are elliptic, let $A$ be a particular position of the variable point $X$. Observe that the product $X \mathbin{\overline{\wedge}} X^{IJ}$ relates the interval $\overline{AA^I}/A^{IJ}$ to the interior interval $\overline{A^{IJ}A^J}/A$; then use Axiom 3·51.

5. If the harmonic relations H($BC$, $AA'$), H($CA$, $BB'$), H($AB$, $CC'$) all hold, prove that the pairs $AA'$, $BB'$, $CC'$ belong to an involution.* *Hint:* Apply 4·63 to the involution $BCAA' \mathbin{\overline{\wedge}} ACBB'$, and deduce $ABCC' \mathbin{\overline{\wedge}} A'B'C'C$.

**4·7  Quadrangular Set of Six Points.**  When we say that four collinear points $A$, $A'$, $B$, $B'$ determine an involution $(AA')(BB')$, we mean that, for any given point $X$ on the line, we can find a companion $X'$ such that $XX'$ is a pair of the involution. So far our only method for constructing $X$ is by applying, to the special case of $AA'B \mathbin{\overline{\wedge}} A'AB'$, the general procedure for $ABC \mathbin{\overline{\wedge}} A'B'C'$ (on one line) described on

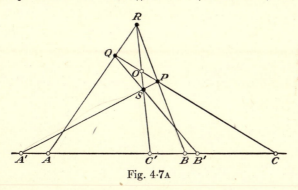

Fig. 4·7A

page 39. This would require the use of nine auxiliary points outside the line. A far more elegant procedure, using only four auxiliary points, is suggested by the following important theorem (due to Pappus):

**4·71**  *The three pairs of opposite sides of a quadrangle meet any line (not through a vertex†) in three pairs of an involution.*

*Proof:* The given line cannot pass through more than two diagonal points; therefore let us assume that it does not pass through $0 = PQ \cdot RS$. Let $A$, $A'$, $B$, $B'$, $C$, $C'$ be the sections of the respective sides $QR$, $PS$, $RP$, $QS$, $PQ$, $RS$, as in Fig. 4·7A. Then $C \neq C'$, and

$$ABCC' \mathbin{\overset{R}{\overline{\wedge}}} QPCO \mathbin{\overset{S}{\overline{\wedge}}} B'A'CC'$$

The desired conclusion now follows from 4·66.

* Mathews (Ref. 25, p. 88).
  † von Staudt (Ref. 40, p. 122, §222). If the given line did pass through a vertex, say $S$, we should have the "degenerate involution" $(AS)(BS)$.

We naturally call this the *quadrangular involution* determined on the given line by the quadrangle $PQRS$, and we call $AA'BB'CC'$ a *quadrangular set* of six points.

Conversely, the companion of any given point $C$ in a given involution $(AA')(BB')$ may be constructed as follows: Draw any triangle $PQR$ whose sides $QR$, $RP$, $PQ$ pass through $A$, $B$, $C$, respectively. This determines a quadrangle $PQRS$, where $S = A'P \cdot B'Q$. Five sides pass through the five given points $A$, $A'$, $B$, $B'$, $C$; therefore the remaining side $RS$ determines the desired companion $C' = RS \cdot AB$.

The construction for a harmonic conjugate (Fig. 2·5A) arises as the special case when the line of section joins two diagonal points of the quadrangle, *i.e.*, when $A = A'$ and $B = B'$. We see now that an elliptic involution is just as easy to construct as a hyperbolic involution with given invariant points.

In Fig. 4·4A we used a quadrangle $A_0B_0RS$ to construct the second invariant point $N$ of a hyperbolic projectivity $MAB \overline{\wedge} MA'B'$. The pairs of opposite sides of that quadrangle meet the line $AB$ in the point pairs $MN$, $AB'$, $BA'$. Thus 4·66 and 4·67 remain valid when $N$ coincides with $M$ (or $C'$ with $C$), *i.e.*, when the projectivity is parabolic instead of hyperbolic. Taking 4·45 into consideration, we see that 4·68 likewise remains valid when $M$ and $N$ coincide. In other words:

**4·72**   *M is an invariant point of the involution* $(AB')(BA')$ *if and only if* $MMAB \overline{\wedge} MMA'B'$, *in which case we have also*

$$MMAA' \overline{\wedge} MMBB'.$$

**4·73**   *If M is an invariant point of each of the involutions*

$$(AA_1)(BB_1) \text{ and } (A'A_1)(B'B_1),$$

*it is also an invariant point of* $(AB')(BA')$.

### EXERCISES

1. Take five collinear points $A$, $A'$, $B$, $B'$, $C$, and construct the sixth point of the quadrangular set, as in Fig. 4·7A. Then (below the line, as in Fig. 2·5C) make the analogous construction using $C'$ instead of $C$. Observe that the new line $RS$ passes through $C$.

2. Deduce 4·43 from 4·72.

3. How many parabolic projectivities can be found to relate two given points $A$ and $B$ to two given points $A'$ and $B'$? (None if $AB'//BA'$; two otherwise. See 4·72 and 4·65.)

**4·8   Projective Pencils.**   For simplicity we have considered projective ranges; but all our results can be dualized to give properties of projective *pencils*. For instance, the fundamental theorem 4·21 dualizes as follows:

*A projectivity (between two pencils) is determined when three lines of one pencil and the corresponding three lines of the other are given.*

When the two pencils have distinct centers, we can dualize Fig. 4·2A to obtain the following construction for the line $x'$ related to a given line $x$ in a given projectivity

$$abc \; \overline{\wedge} \; a'b'c'$$

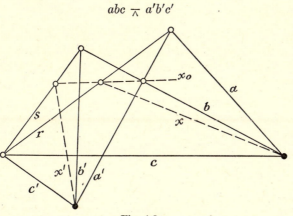

Fig. 4·8A

Draw, as in Fig. 4·8A, the lines

$$r = (a \cdot a')(c \cdot c'), \qquad s = (b \cdot b')(c \cdot c'),$$
$$x_0 = (r \cdot x)(b \cdot a'), \qquad x' = (s \cdot x_0)(a' \cdot b')$$

Again, the dual of 4·23 is as follows:

*If a projectivity between pencils with distinct centers has an invariant line, it is merely a perspectivity* (in the sense of Fig. 2·7B).

### EXERCISES

1. Given five concurrent lines $m$, $a$, $b$, $a'$, $b'$, construct the second invariant line of the hyperbolic projectivity $mab \; \overline{\wedge} \; ma'b'$. (*Hint:* Dualize Fig. 4·4A.)

2. Dualize 4·71. Hence construct the companion of a given line $c$ in a given involution $(aa')(bb')$.

# CHAPTER 5

## TWO-DIMENSIONAL PROJECTIVITIES

We shall find that the one-dimensional projectivity considered in Chap. 4 has two different analogues in two dimensions: one relating points to points and lines to lines, the other relating points to lines and lines to points. The former kind is a collineation, the latter a correlation. Although the general theory is due to von Staudt,[*] and the names *collineation* and *correlation* to Möbius (1827), some special collineations were used much earlier, *e.g.*, by Newton and La Hire.[†] Moreover, the classical transformations of the Euclidean plane, *viz.*, translations, rotations, reflections, and dilatations, all provide instances of collineations. Poncelet considered the relation between the central projections of a plane figure onto another plane from two different centers. He called this special collineation a *homology*. In Sec. 5·2 we shall give a purely two-dimensional account of it. Poncelet also considered a special correlation: the polarity induced by a conic. In Sec. 5·5, following von Staudt again, we obtain the same transformation without using a conic. We then find that several famous properties of conics are really properties of polarities (which are simply correlations of period two).

**5·1 Collineation.**[‡]   We recall that a collineation is a point-to-point correspondence preserving collinearity and consequently preserving the harmonic relation. Thus a collineation induces a projectivity between ranges on corresponding lines, and a projectivity between pencils through corresponding points.

**5·11** *If a quadrilateral or a quadrangle is invariant, the collineation can only be the identity.*

*Proof:* Suppose a quadrilateral is invariant. Then the sides are four invariant lines, and the vertices (where the sides meet in pairs) are six

---

[*] Ref. 40, pp. 60–66, 125–136.

[†] See Coolidge (Ref. 5, p. 47).

[‡] von Staudt (Ref. 40, pp. 61, 66, §§123, 130); Cremona (Ref. 8, p. 78); Enriques (Ref. 11, p. 159).

invariant points, three on each side. Hence, by 4·12, every point on each side is invariant. Any other line contains invariant points where it meets the sides and is consequently invariant. Thus the collineation must be the identity. The dual argument gives the same result for a quadrangle.

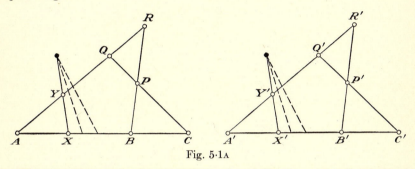

Fig. 5·1A

The fundamental theorem has the following two-dimensional analogue:

**5·12** *A collineation is determined when two corresponding quadri-laterals (or quadrangles) are given.*

*Proof:* Let $ABCPQR$ and $A'B'C'P'Q'R'$ be the two given quadri-laterals. A line of general position may be described as $XY$, with $X$ on $AB$ and $Y$ on $AQ$, as in Fig. 5·1A. This determines a line $X'Y'$, where

$$ABCX \;\overline{\wedge}\; A'B'C'X' \qquad \text{and} \qquad AQRY \;\overline{\wedge}\; A'Q'R'Y'$$

To prove that the correspondence $XY \to X'Y'$ is a collineation, we have to verify that concurrent lines correspond to concurrent lines, i.e., that a pencil of lines $XY$ leads to a pencil of lines $X'Y'$. (It will then follow that collinear points correspond to collinear points.)

For this purpose, let $XY$ vary in a pencil, so that $X \;\overline{\overline{\wedge}}\; Y$. By our definition of $X'Y'$ we now have

$$X' \;\overline{\wedge}\; X \;\overline{\overline{\wedge}}\; Y \;\overline{\wedge}\; Y'$$

Since $A$ is the invariant point of the perspectivity $X \;\overline{\overline{\wedge}}\; Y$, $A'$ must be an invariant point of the projectivity $X' \;\overline{\wedge}\; Y'$. Hence, by 4·23, this projectivity is again a perspectivity, and $X'Y'$ varies in a pencil, as desired.

Finally, the collineation $ABCPQR \to A'B'C'P'Q'R'$ is unique, by 5·11.

### EXERCISE

Give two reasons why inversion wo a circle is *not* a collineation.

**5·2    Perspective Collineation.**    In particular, a collineation having two invariant points $M$ and $N$ may be described as relating a quadrangle $MNAB$ to a quadrangle $MNA'B'$. It may happen that the two corresponding points $MN \cdot AB$ and $MN \cdot A'B'$ coincide, as in Fig. 5·2A. Then the line $o = MN$ contains three invariant points and consequently consists entirely of invariant points. Thus $AA'$ and $BB'$, meeting $o$ in invariant points, are invariant lines and intersect in an invariant point $O$.

Every line through $O$ is invariant. For if $O$ does not lie on $o$, such a line joins $O$ to an invariant point on $o$; and if $O$ does lie on $o$, we have

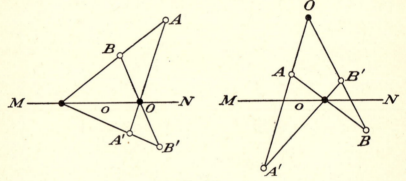

Fig. 5·2A

three invariant lines through it, namely, $o$, $AA'$, $BB'$. Such a *perspective* collineation, leaving invariant every line through a certain point $O$ and every point on a certain line $o$, is called an *elation* or a *homology* according as the center $O$ and axis $o$ are or are not incident.

The above remarks show that every collineation that has three collinear invariant points (or three concurrent invariant lines) is either an elation or a homology.

**5·21**    *An elation or a homology is determined when its center and axis and one pair of corresponding points (collinear with the center) are given.*[*]

*Proof:* Let $AA'$ be the given pair, collinear with the center $O$. Any point $X$ (not on $OA$) determines $C = AX \cdot o$ and $X' = OX \cdot CA'$, as in Fig. 5·2B. Since all points on $o$ and all lines through $O$ are invariant, the collineation must relate $X = OX \cdot CA$ to the point $X'$ so defined.

---

[*] Veblen and Young (Ref. 42, p. 72). This seems to be the first appearance of the word *elation*. Poncelet (Ref. 29, pp. 155–169) called every perspective collineation a homology, and Enriques (Ref. 11, p. 163) distinguished the elation as a "special" homology.

**5·22**  Corollary: *An elation is determined when its axis and one pair of corresponding points are given.*

The relation between corresponding points on a given line through $O$ is

$$X \overset{A}{\barwedge} C \overset{A'}{\barwedge} X'$$

The only possible invariant points of this projectivity $X \barwedge X'$ are on $AA'$ or $o$. Hence:

**5·23**  *An elation or a homology induces a parabolic or hyperbolic projectivity (respectively) on any line through its center.*

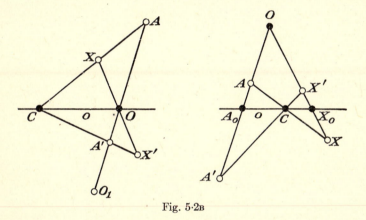

Fig. 5·2ᴮ

Turning to Fig. 2·2ᴀ, we observe that the homology that takes $P$ to $P'$ (with center $O$ and axis $ABC$) also takes $Q$ to $Q'$ and $R$ to $R'$. In the special case when $O$ lies on $ABC$, we have an elation instead of a homology. Hence:

**5·24**  *Any pair of Desargues triangles are related by a homology or an elation.*

All the invariant points of an elation lie on its axis. Conversely, a collineation that has a line of invariant points and no others can only be an elation. These remarks will enable us to prove the following:

**5·25**  *The product of two elations having the same axis is another such elation (if it is not merely the identity).*

*Proof:* Clearly, each point on the axis is invariant. If any other point $A$ were invariant, too, the first elation would take $A$ to some different point $A'$ and the second would take $A'$ back to $A$. By 5·22 the second would then be just the inverse of the first. Hence, apart from that trivial case, all the invariant points of the product must lie on the axis (cf. 4·45).

Consider once more the homology determined by $O$, $o$, $A$, and $A'$, as in the second part of Fig. 5·2B. Let $OA$ and $OX$ meet $o$ in $A_0$ and $X_0$. The homology is said to be *harmonic* if $H(OA_0, AA')$. In this case we can simply locate $X'$ as the harmonic conjugate of $X$ wo $O$ and $X_0$. Hence:

**5·26**  *A harmonic homology is determined when its center and axis are given.*

By an argument similar to that used in proving 5·25, we have the following:

**5·27**  *The product of two harmonic homologies having the same axis is an elation.*

Conversely:

**5·28**  *An elation with axis o may be expressed as the product of two harmonic homologies having this same axis o.*

*Proof:* Let the elation be determined by $o$, $A$, and $A'$, as in the first part of Fig. 5·2B, and let $O_1$ be the harmonic conjugate of $O = AA' \cdot o$ wo $A$ and $A'$. Then the harmonic homologies with centers $A$ and $O_1$

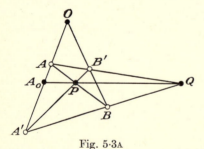

will have the desired effect, since the first leaves $A$ invariant, while the second takes it to $A'$.

Fig. 5·3A

### EXERCISES

1. Show that the central projections of a plane figure onto another plane from two different centers are related by an elation or homology. (*Hint:* The relation is a collineation with a line of invariant points where the two planes intersect.)

2. Justify the above statement that a collineation having a line of invariant points and no others can only be an elation.

**5.3 Involutory Collineation.**  A collineation may be *periodic* according to the definition in Sec. 4·5. A collineation of period 2 is said to be *involutory*.

**5·31**  *Every involutory collineation is a harmonic homology.*

*Proof:* Let the given involutory collineation interchange the pair of points $AA'$ and also another pair $BB'$ (not on the line $AA'$). The invariant lines $AA'$ and $BB'$ intersect in an invariant point $O$, as in

Fig. 5·3a. Since the collineation interchanges the pair of lines $AB$, $A'B'$ and likewise $AB'$, $A'B$, the two points

$$P = AB \cdot A'B' \qquad \text{and} \qquad Q = AB' \cdot A'B$$

are invariant. Moreover, the two invariant lines $AA'$ and $PQ$ meet in a third invariant point $A_0$ on $PQ$. Hence the collineation is either an elation or a homology. Since the invariant point $O$ does not lie on $PQ$, it must be a homology; and since $H(AA', OA_0)$ it is a *harmonic homology*.

**5·32**　*Two harmonic homologies commute if and only if the center of each lies on the axis of the other.*

*Proof:* Let two harmonic homologies $H$, $H'$ have centers $O$, $O'$ and axes $o$, $o'$. If $O$ lies on $o'$ and $O'$ on $o$, any two points that are harmonic conjugates wo $O$ and $O'$ are interchanged by each homology. Hence the product $HH'$ leaves invariant every point on $OO'$, and similarly every line through $o \cdot o'$. Thus $HH'$ is a homology. To see that it is harmonic, we consider two points that are harmonic conjugates wo $O$ and $o \cdot o'$. These are interchanged by $H$ but invariant for $H'$. Thus $HH'$, and similarly $H'H$, is the harmonic homology with center $o \cdot o'$ and axis $OO'$.

Conversely, if $H$ and $H'$ commute, their product $HH'$, being equal to its inverse $H'H$, is an involutory collineation, *i.e.*, another harmonic homology. Now, $H'$ transforms any point $X$ on $o$ into a point $X^{H'}$ on $o^{H'}$. Since $X$ is invariant for $H$, the point

$$X^{H'} = X^{HH'} = X^{H'H}$$

is likewise invariant for $H$. But $X^{H'}$ may be *any* point on $o^{H'}$. Therefore $o^{H'}$ coincides with $o$, the axis of $H$; that is, $o$ is invariant for $H'$ and either coincides with $o'$ or passes through $O'$. The former possibility is ruled out since, by 5·27, the product $HH'$ would then be an elation. Hence $o$ passes through $O'$, and similarly $o'$ through $O$.

### EXERCISE

Show that the product of three harmonic homologies, whose centers and axes are the vertices and sides of a triangle, is the identity.

**5·4　Correlation.** We come now to the second kind of two-dimensional projectivity. A *correlation* is a point-to-line correspondence relating collinear points to concurrent lines; it therefore relates concurrent lines to collinear points. Incidences are dualized: we have $X \to x'$ and $x \to X'$, where line $x'$ passes through point $X'$ if and only if point $X$ lies on line $x$. The range of points $X$ on a given line $x$ corresponds to a pencil of lines $x'$ through the corresponding point $X'$. Since a quadrangle corresponds to a quadrilateral, four positions of $X$ forming a harmonic set of points correspond to four positions of $x'$ forming a harmonic set of lines. Thus a correlation induces a projectivity between any range and the corresponding pencil.

**5·41**   *A correlation is determined when a quadrilateral and the corresponding quadrangle are given.*[*]

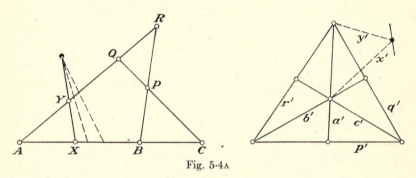

Fig. 5·4A

*Proof:* Let $ABCPQR$ and $a'b'c'p'q'r'$ be the given quadrilateral and quadrangle, as in Fig. 5·4A. A line $XY$, with $X$ on $AB$ and $Y$ on $AQ$, determines a point $x' \cdot y'$ where

$$ABCX \overline{\wedge} a'b'c'x' \qquad \text{and} \qquad AQRY \overline{\wedge} a'q'r'y'$$

To prove that the correspondence $XY \to x' \cdot y'$ is a correlation, we have to verify that a pencil of lines $XY$ leads to a range of points $x' \cdot y'$.

Let $XY$ vary in a pencil, so that $X \overline{\wedge} Y$ and therefore $x' \overline{\wedge} y'$. Since $A$ is the invariant point of the perspectivity $X \overline{\wedge} Y$, $a'$ must be an invariant line of the projectivity $x' \overline{\wedge} y'$. Hence, by the dual of 4·23, this projectivity is a perspectivity, and $x' \cdot y'$ varies in a range as desired.

Since the product of two correlations is a collineation, the uniqueness of the correlation $ABCPQR \to a'b'c'p'q'r'$ is another consequence of 5·11. In fact, by combining one such correlation with the inverse of another, we should obtain a collineation leaving the quadrilateral invariant.

**5·5   Polarity.**   In general, a correlation relates a point $X$ to a line $x'$ and relates this line to a new point $X''$. The correlation is involutory

* Veblen and Young (Ref. 42, p. 264).

(*i.e.*, of period 2) if $X''$ always coincides with $X$, in which case we may omit the prime ['] without causing any confusion. An involutory correlation is called a *polarity*. Thus a polarity is a correlation that relates $X$ to $x$, and vice versa. Following Servois and Gergonne, we call $X$ the *pole* of $x$ and $x$ the *polar* of $X$.

This terminology may be justified as follows: The section of a sphere by a plane through the center is a great circle, and the "axis" of this circle meets the sphere in two *poles* (*e.g.*, the North Pole and South Pole are the poles of the equator). When we make a gnomonic projection (from the center onto an arbitrary plane), the great circle and the two poles yield a straight line and a single point, the *pole* of the line. This is easily seen to satisfy the above description of a polarity. (It differs from the polarity wo a conic, in that no point lies on its own polar.)

As a consequence of the general properties of a correlation, we see that the polars of all the points on a line $a$ form a projectively related pencil of lines through the pole $A$.

Since a polarity dualizes incidences, if $A$ lies on $b$, $a$ passes through $B$. In this case we say that $A$ and $B$ are *conjugate points*, $a$ and $b$ are *conjugate lines*. If $A$ and $a$ are incident, $A$ is a self-conjugate point and $a$ a self-conjugate line. So far as we can tell at present, it might happen that every point and line would be self-conjugate; but the following theorem shows that this is not so:[*]

**5·51**  *The join of two self-conjugate points cannot be a self-conjugate line.*

*Proof:* If a line $s$ contains two self-conjugate points $A$ and $B$, the polars of these points are $a = AS$ and $b = BS$, where $S$ is the pole of $s$. Since a polarity is a one-to-one correspondence between points and lines, these two polars must be distinct; therefore $S$ does not lie on $AB$, that is, $s$ is not self-conjugate.

As a further limitation on the occurrence of self-conjugacy, we shall prove the following:[*]

**5·52**  *It is impossible for a line to contain more than two self-conjugate points.*

*Proof:* Let $A$ and $B$ be two self-conjugate points on $s$, and let $P$ be a point on $AS$ or $a$, distinct from $A$ and $S$. Let the polar $p$ meet $b$ in $Q$ (see Fig. 5·5A). Then $Q = b \cdot p$ is the pole of $BP = q$, which meets $p$ in $R$, say. Also, $R = p \cdot q$ is the pole of $PQ = r$, which meets $s$ in $C$, say. Finally, $C = r \cdot s$ is the pole of $RS = c$, which meets $s$ in $D$, the harmonic conjugate of $C$ wo $A$ and $B$. Now, $C$ cannot coincide with $A$ or $B$; for then $P$ would coincide with $A$ or $S$. Hence $C$, not lying on $c$, is not self-conjugate.

We thus have, on $s$, two self-conjugate points $A$, $B$ and a non-self-conjugate point $C$. But the self-conjugate points on $s$ are the invariant points of the projectivity $X \barwedge x \cdot s$ induced on $s$ by the given polarity. Hence this projectivity is not the identity, and cannot have more than

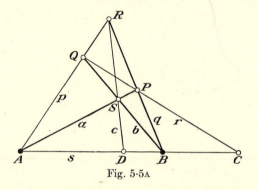

Fig. 5·5A

two invariant points; *i.e.*, the line $s$ cannot contain more than two self-conjugate points of the polarity.

We can now easily prove:

**5·53**  *A polarity induces an involution of conjugate points on any line that is not self-conjugate and an involution of conjugate lines through any point that is not self-conjugate.\**

*Proof:* The projectivity $X \barwedge x \cdot s$ on $s$ relates any non-self-conjugate point $C$ to another point $D = c \cdot s$, whose polar is $CS$. Hence the same projectivity relates $D$ to $C$; that is, it *interchanges* $C$ and $D$. By 4·61, it must be an involution. (In the proof of 5·52 this was a hyperbolic involution; but it can just as well be elliptic.)

Dually, $x$ and $XS$ are paired in the involution of conjugate lines through $S$.

The following theorem is famous (as a property of a conic):

**5·54**  **Hesse's Theorem:** *If two pairs of opposite vertices of a quadrilateral are pairs of conjugate points (in a given polarity), then the third pair of opposite vertices is likewise a pair of conjugate points.*

---

\* von Staudt (Ref. 40, p. 134, §239). On a self-conjugate line $s$ the conjugate points form a "degenerate involution": each point is conjugate to the single point $S$.

Instead of this, we shall prove the dual:

**5·55** *If two pairs of opposite sides of a quadrangle are pairs of conjugate lines, then the third pair of opposite sides is likewise a pair of conjugate lines.*\*

*Proof:* Given $QR$ conjugate to $PS$, and $PR$ conjugate to $QS$, we wish to show that $PQ$ is conjugate to $RS$. Let these six lines meet $s$ (the polar of $S$) in $A$, $A'$, $B$, $B'$, $C$, $C'$, as in Fig. 5·5B (or Fig. 4·7A). Since $PS$ passes through $S$ and is conjugate to $QR$, its pole is $s \cdot QR = A$; hence $PS$ is $a$. Similarly $QS$ is $b$. Thus the involution of conjugate points on

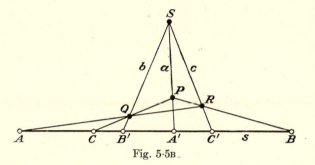

Fig. 5·5B

$s$ is $(AA')(BB')$. By 4·71, $CC'$ is another pair of this involution. Hence $RSC'$ is the polar of $C$ and is conjugate to $PQC$.

The above proof would break down if the point $S$ were self-conjugate. For the sake of completeness, therefore, we must make sure that the four vertices of such a quadrangle cannot all be self-conjugate.

**5·56** *If all four vertices of a quadrangle are self-conjugate points, at most one pair of opposite sides can be conjugate lines.*

*Proof*†: Let $PQRS$ be a quadrangle of self-conjugate points, so that $S$ lies on its own polar $s$ as well as on the sides $a = PS$, $b = QS$, $c = RS$, whose poles are $A = p \cdot s$, $B = q \cdot s$, $C = r \cdot s$. If the sides $QR$ and $PS$ are conjugate lines, let $O$ be their point of intersection, as in Fig. 5·5c. Since the pole $A$ lies on $QR$, and $q \cdot r$ on $a$, we have

$$asbc \underset{\wedge}{\overline{\phantom{x}}} ASBC \overset{q \cdot r}{\underset{\wedge}{=}} AOQR \underset{\wedge}{\overline{\phantom{x}}} sabc,$$

whence $H(sa,bc)$. Similarly, if $RP$ were conjugate to $QS$ we would have $H(sb,ca)$, and if $PQ$ were conjugate to $RS$ we would have $H(sc,ab)$. By 3·21 and 3·17, only one of these three harmonic relations can hold.

\* von Staudt (Ref. 40, p. 136, §244). For a slightly different proof see Cremona (Ref. 8, p. 238).

† Kindly supplied by Patrick Du Val of Istanbul University.

## EXERCISES

**1.** Observe that the polarity of 5·52 (Fig. 5·5A) relates the quadrangle *PQRS* to the quadrilateral *ABCPQR* and the harmonic set of points *A*, *B*, *C*, *D* to the harmonic set of lines *SA*, *SB*, *SD*, *SC*.

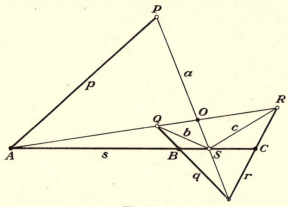

Fig. 5·5c

**2.** Justify the statement (in the proof of 5·52) that *C* coinciding with *A* or *B* would make *P* coincide with *A* or *S*.

**5·6   Polar and Self-polar Triangles.**\* The *polar triangle* of a given triangle is formed by the polars of the vertices and the poles of the sides.

**5·61**   Chasles's Theorem: *A triangle and its polar triangle (if distinct) are a pair of Desargues triangles.*

*Proof:* Let *P′Q′R′* be the polar triangle of *PQR*. Consider the quadrangle *PQRS*, where *S* is *PP′* · *QQ′*. We have *PS* (through *P′*) conjugate to *p′* = *QR*, and *QS* (through *Q′*) conjugate to *q′* = *PR*. By 5·55, *RS* is conjugate to *r′* = *PQ* and therefore passes through *R′*, as in Fig. 5·6A. Thus corresponding vertices of the two triangles *PQR* and *P′Q′R′* are collinear with *S*.

A triangle is said to be *self-polar* if each vertex is the pole of the opposite side. Any two points (or lines) that are conjugate but not self-conjugate determine a self-polar triangle; for if *A* and *B* are conjugate points on a non-self-conjugate line *c*, each vertex of triangle *ABC* is the pole of the opposite side. Any two vertices (or sides) are conjugate. The occurrence of such a triangle is characteristic of a polarity, as the following theorem shows:

\* von Staudt (Ref. 40, pp. 131–135, §§234–242); Enriques (Ref. 11, pp. 182, 187).

**5·62**  *Any correlation that relates the three vertices of one triangle to the respectively opposite sides is a polarity.*

*Proof:* Consider the correlation $ABCP \rightarrow abcp$, where $a$, $b$, $c$ are the sides of the given triangle $ABC$ and $P$ is a point of general position. Let $A_p$, $B_p$, $C_p$, $P_a$, $P_b$, $P_c$ denote the respective points

$$a \cdot p, \qquad b \cdot p, \qquad c \cdot p, \qquad a \cdot AP, \qquad b \cdot BP, \qquad c \cdot CP$$

as in Fig. 5·6B. The correlation not only relates $A$, $B$, $C$ to $a$, $b$, $c$ but

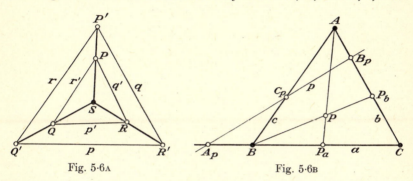

| Fig. 5·6A | Fig. 5·6B |

also relates $a$, $b$, $c$ to $A$, $B$, $C$; for it relates $a = BC$ to $A = b \cdot c$, and so on. Moreover, it relates $AP$ to $A_p$, and consequently $P_a$ to $AA_p$. We wish to show that it relates $p$ to $P$.

Consider the projectivity $X \overline{\wedge} x \cdot a$ induced on $a$. Since $b \cdot a$ and $c \cdot a$ are $C$ and $B$, this projectivity interchanges $B$ and $C$; hence it is an involution. Since $P_aA_p$ is a pair of the involution, the correlation works as follows: $A_p \rightarrow AP_a$; similarly $B_p \rightarrow BP_b$; hence $A_pB_p \rightarrow AP_a \cdot BP_b$; that is, $p \rightarrow P$.

**5·63**  Corollary: *A polarity is determined when a self-polar triangle and one further pole and polar are given.*

*Notation:* The polarity with self-polar triangle $ABC$, relating $P$ and $p$, is denoted by

$$(ABC)(Pp)$$

In using this symbol, we assume that $P$ does not lie on a side of triangle $ABC$ and that $p$ does not pass through a vertex.

We proceed to describe a construction (Fig. 5·6c) for the polar of any point $X$ in a given polarity $(ABC)(Pp)$. If $X$ is distinct from $P$, it cannot lie on more than one of the lines $AP$, $BP$, $CP$, and we may suppose $A$, $B$, $C$ to be named in such an order that $X$ does not lie on either of $AP$, $BP$. The construction is as follows:

**5·64**  *The polar of any point $X$ in the polarity $(ABC)(Pp)$ is the line*
$[AP \cdot (a \cdot PX)(p \cdot AX)][BP \cdot (b \cdot PX)(p \cdot BX)]$.

*Proof:* We have to show that the polar of $X$ is the line $YZ$ determined by

$$A_1 = a \cdot PX, \quad E = p \cdot AX, \quad Y = AP \cdot A_1F$$
$$B_1 = b \cdot PX, \quad F = p \cdot BX, \quad Z = BP \cdot B_1F$$

By Hesse's theorem (our 5·54), applied to the quadrilateral $AA_1PEXY$ (Fig. 5·6c), since $AA_1$ and $PE$ are pairs of conjugate points, $XY$ is another pair of conjugate points. Thus $X$ is conjugate to $Y$ and similarly to $Z$. Therefore $x = YZ$.

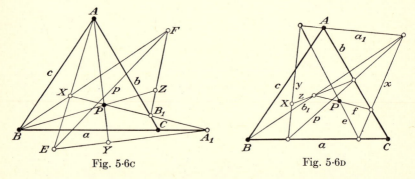

Fig. 5·6c             Fig. 5·6D

In 5·63 a polarity was determined by means of a self-polar triangle and one further pole and polar. Another way would be by means of a *self-polar pentagon* such as $PBC_pB_pC$, that is, a pentagon whose vertices are the poles of the respectively opposite sides. In fact:

**5·65**  *The correlation that relates four vertices of a pentagon to the respectively opposite sides relates the fifth vertex to the fifth side and is a polarity.*

*Proof:* The correlation that relates the vertices $Q$, $R$, $S$, $T$ of a pentagon $PQRST$ to the sides $ST$, $TP$, $PQ$, $QR$ also relates the point $A = QR \cdot ST$ to the line $(ST \cdot TP)(PQ \cdot QR) = TQ$. Thus it relates each vertex of triangle $AQT$ to the opposite side and is a polarity, by 5·62. Finally, since it relates $RS$ to $TP \cdot PQ = P$, it also relates $P$ to $RS$.

### EXERCISES

1. Let $AA$, $BB'$, $CC'$ be a quadrangular set of points and $S$ any point outside their line. Prove that any correlation which relates $S$, $A$, $B$, $C$ to $AB$, $SA'$, $SB'$, $SC'$ is a polarity (Fig. 5·5B).

2. Show how the construction 5·64 becomes partly indeterminate when $P$ is given on $BC$ or $AB$ (and $p$ through $A$ or $C$, respectively). *Hint:* If $P$ lies on $BC$,

so also does $Z$; $Y$ may be any point on $AP$, and $x$ any line through $Z$. If $P$ lies on $AB$, $Y$ and $Z$ coincide; for then $ABY$ is the Pappus line of the hexagon $A_1EXFB_1C$ (see 4·31).

3. By dualizing 5·64, derive $X$ from $x$ (Fig. 5·6D).

4. Consider the self-polar pentagon $PQRST$ of Theorem 5·65. Let an arbitrary line through $P$ meet $ST$ in $U$ and $QR$ in $V$. Prove that $RU$ and $SV$ are conjugate lines. Deduce a construction for the polar of any given point $X$.

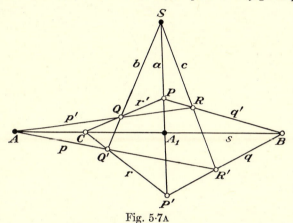

Fig. 5·7A

**5·7   The Self-polarity of the Desargues Configuration.**   Chasles's theorem (our 5·61) has the following interesting converse, due to von Staudt:

**5·71**   *Any pair of Desargues triangles are polar triangles in a certain polarity.*

*Proof:** Let the triangles $PQR$ and $P'Q'R'$ be "in perspective" from the point $S$ and the line $ABC$, as in Fig. 5·7A. Consider the polarity $(SAA_1)(Q'q')$, where $A_1 = AB \cdot SP$ and $q' = RP$. Since $QR$ (through $A$) is conjugate to $PS(= SA_1)$, while $QS$ (through $Q'$) is conjugate to $RP$ ($= q'$), 5·55 shows that $RS$ is conjugate to $PQ$. Thus $QS$ ($= Q'S$) is the polar of $q' \cdot s = B$, and $RS$ is the polar of $PQ \cdot s = C$. The polarity works as follows: The polars of

$$Q', S, A, B, C, P, R, P', Q, R'$$

are

| | | | | |
|---|---|---|---|---|
| $q' = RP,$ | $s = AA_1,$ | $a = PS,$ | $b = QS,$ | $c = RS$ |
| $p = AQ',$ | $r = CQ',$ | $p' = AR,$ | $q = BP',$ | $r' = CP$ |

(Each polar in the second row is derived from two previous results; for example, $p$ is $AQ'$ because $P$ is $a \cdot q'$.) In this manner we see that the

---

* This is perhaps a little simpler than von Staudt (Ref. 40, p. 134, §241).

10 lines of the Desargues configuration are the polars of its 10 points. In particular, $PQR$ and $P'Q'R'$ are polar triangles.

## EXERCISES

1. Let $\Pi$ denote the polarity of 5·71 and $\Gamma$ the homology or elation of 5·24. Prove that $\Gamma\Pi = (PQR)(Ss)$ and $\Pi\Gamma = (P'Q'R')(Ss)$. Deduce that these two polarities coincide when $\Gamma$ is a harmonic homology.

2. Prove that the polarity $(ABC')(Nn)$ interchanges the nine points and nine lines of the Pappus configuration (Fig. 4·3A and B) if and only if the lines $LA'$, $MB'$, $CN$ are concurrent. *Hint:* If those three lines are concurrent, consider the polarity of 5·71 as applied to triangles $LMC$ and $A'B'N$. Show that $B$ is the pole of $AC'$, by applying 5·55 to the quadrangle $AC'BB'$.* (Since the general Pappus configuration is self-dual without being self-polar, the old controversy between Poncelet and Gergonne is settled in the latter's favor.)

3. Show that the Desargues configuration contains 12 self-polar pentagons such as $ABPSQ'$, any one of which could be used to determine von Staudt's polarity (cf. Sec. 2·2, Exercise 4).

**5·8   Pencil and Range of Polarities.**† The various polarities that have a given self-polar triangle and a given pair of conjugate points (or lines) are said to form a pencil (or range) of polarities. Thus, given a self-polar triangle $ABC$, the polarities $(ABC)(Pp)$ form a *pencil* if $P$ is fixed while $p$ varies in a pencil of lines; dually, they form a *range* if $p$ is fixed while $P$ varies along a line. Thus any two polarities that have a common self-polar triangle belong to a definite pencil and to a definite range.

**5·81**  *The polars of any fixed point $X$, wo a pencil of polarities, form a pencil of lines.*

*Proof:* Referring to the construction 5·64 for the polar of $X$ in $(ABC)(Pp)$, let $p$ rotate about a fixed point $P'$. Then $Y$ varies on $AP$, and $Z$ on $BP$, in such a way that

$$ Y \stackrel{A_1}{\overline{\wedge}} E \stackrel{P'}{\overline{\wedge}} F \stackrel{B_1}{\overline{\wedge}} Z $$

But the projectivity $Y \overline{\wedge} Z$ has $P$ as an invariant point, arising when $p$ is $P'X$ (so that $E$ and $F$ coincide with $X$). Hence $Y \overline{\overline{\wedge}} Z$; that is, the line $x = YZ$ passes through a fixed point $X'$.

An important special case occurs when $P'$ lies on a side of the triangle, say on $a$, so that the fixed point $P'$ is the $A_p$ of Fig. 5·6B. Then the

---

* This method is due to Alex Rosenberg, an undergraduate at the University of Toronto.

† This section may be omitted on first reading or in a short course.

involution of conjugate points on $a$ is $(BC)(A_pP_a)$, the same for all the polarities. The self-polar triangle is no longer unique; for any pair of this involution would form with $A$ a self-polar triangle that could be used instead of $ABC$. Any point $X$ projects from $A$ into a point $X_a$ on $a$, and the polars $x$ form a pencil of lines through $A_x$, the companion of $X_a$ in the involution $(BC)(A_pP_a)$. Conversely, if any line $x$ meets $a$ in $A_x$, its poles $X$ all lie on the fixed line $AX_a$, where $X_a$ is the companion of $A_x$. Hence this kind of pencil of polarities is at the same time a range. Let us simply call it a *self-dual system* of polarities. To sum up:

**5·82**  *A self-dual system admits a line $a$ on which the involution of conjugate points is the same for all the polarities. The polars of any point $P$ form a pencil of lines through a fixed point $A_p$ on $a$, and the poles of any line $p$ form a range on a fixed line $AP_a$, where $A$ is the pole of $a$ (for all the polarities).*

The product of two polarities (or, indeed, of any two correlations) is a collineation. In particular,

**5·83**  *If two polarities belong to a self-dual system, their product is a homology.*

*Proof:* Since the polarities induce the same involution of conjugate points on the line $a$ and the same involution of conjugate lines through the point $A$, their product leaves invariant every point on $a$ and every line through $A$.

Conversely,

**5·84**  *Any homology\* can be expressed as the product of two polarities belonging to a self-dual system.*

*Proof:* The homology with center $A$ and axis $BC$, relating $P_1$ to $P_2$ (on a line through $A$), is the product of polarities $(ABC)(P_1p)$ and $(ABC)(P_2p)$, where $p$ is an arbitrary line.

### EXERCISES

1. Show that the relation between $X$ and $X'$ in 5·81 is symmetric (*i.e.*, involutory): the polars of $X'$ all pass through $X$. (*Hint:* $X$ and $X'$ are conjugate points.)

2. Show that the correspondence $X \to X'$ is not a collineation. (*Hint:* Take $X$ at various positions on $AB$.)

3. Dualize 5·81.

---

\* Veblen and Young (Ref. 42, p. 265) prove the more general statement that any collineation can be expressed as the product of two polarities. But their construction breaks down if the collineation is involutory.

4. Show that the two polarities $(ABC)(P_1p)$ and $(ABC)(P_2p)$ will commute if $H(AP_a, P_1P_2)$.

5. Show that the polarities of Sec. 5·6, Exercises 1 and 2, belong to self-dual systems.

6. Show that a range of polarities is determined by a self-polar pentagon $PQRST$, where $P$ varies along a line while the other four vertices remain fixed.

# CHAPTER 6

## CONICS

**6·1 Historical Remarks.** The study of conic sections (or, briefly, conics) is said to have begun in 430 B.C., when the Athenians, suffering from a plague, appealed to the oracle at Delos and were told to double the size of Apollo's cubical altar. Let $a$ denote the edge of the original cube and $x$ that of the enlarged one; then the requirement is

$$x^3 = 2a^3$$

or $x/a = \sqrt[3]{2}$. Hippocrates reduced the problem to that of finding values of $x$ and $y$ to satisfy the equations

$$\frac{a}{x} = \frac{x}{y} = \frac{y}{2a}$$

Menaechmus, about 340 B.C., gave two solutions: one using the two parabolas $y^2 = 2ax$, $x^2 = ay$, and the other using the latter parabola along with the rectangular hyperbola $xy = 2a^2$. Without the benefit of algebraic notation, this was surely a marvellous achievement. In fact, it shows that Menaechmus came close to anticipating by 2000 years the analytic geometry of Fermat and Descartes. He presumably obtained these curves as plane sections of a right circular cone. (Hence the name *conic section*.) But purely two-dimensional constructions were soon devised by his successors. According to Zeuthen, it was Euclid who first constructed a conic as the locus of a point whose distance from a fixed point (focus) is proportional to its distance from a fixed line (directrix). The names *ellipse, parabola, hyperbola* are due to Apollonius (262–200 B.C.), who discovered an astonishing number of their properties. He even anticipated Steiner's theorem (our 6·52). Some further results were obtained by Pappus, about A.D. 300, but after that time the whole subject was forgotten for 12 centuries.

In fact, no new contribution of any importance was made until 1522, when Verner of Nuremberg derived certain properties of conics by projecting a circle. For the next three centuries, apart from Kepler, Newton, Maclaurin, and Braikenridge, the subject was developed largely by Frenchmen: Desargues, Pascal, Mydorge, La Hire, and then Gergonne, Brianchon, Poncelet, Chasles. Kepler showed how a parabola is at once the limiting form of an ellipse and of a hyperbola, thus paving the way for consideration of the general conic. The name of Braikenridge (1700–1759) is not very familiar, but he shares with Maclaurin the honor of discovering the first nonmetrical construction for a conic.

The more recent developments are dominated by the names of two great Germans: Steiner (1796–1863) and von Staudt. We shall see, in Sec. 6·5, how their two ways of approaching the conic may be reconciled. The polarity induced by a conic is implicit in some of the work of Apollonius and was clearly understood by La Hire (1640–1718), but it was von Staudt who turned the tables by allowing the polarity to define the conic. This standpoint provides the most symmetrical definition for conics and emphasizes their self-dual nature, as we shall see in Sec. 6·3.

**6·2 Elliptic and Hyperbolic Polarities.*** We recall that an involution is hyperbolic or elliptic according as it does or does not admit an invariant point, that a hyperbolic involution has not merely one but two invariant points, and that the involution $(AB)(CD)$ is elliptic or hyperbolic according as $D$ does or does not lie in the segment $AB/C$.

Analogously, a polarity is said to be *hyperbolic* or *elliptic* according as it does or does not admit a self-conjugate point, *i.e.*, a point lying on its own polar. We shall find that a hyperbolic polarity has not merely one but infinitely many self-conjugate points, forming a curve (in fact a conic), and that the polarity $(ABC)(Pp)$ is elliptic or hyperbolic according as $P$ does or does not lie in the triangular region $ABC/p$.

**6·21** *If $P$ lies in $ABC/p$, the polarity $(ABC)(Pp)$ is elliptic.*

*Proof:* Let $p$ meet the sides of triangle $ABC$ in $A_p$, $B_p$, $C_p$, as in Fig. 6·2A, and let the three lines $AP$, $BP$, $CP$ meet the respective sides in $P_a$, $P_b$, $P_c$. If $P$ lies in the region $ABC/p$, then $P_a$ lies in the segment $BC/A_p$ and the involution $(BC)(A_pP_a)$ is elliptic. But this is the involution of conjugate points on the line $BC$. Similarly, the involutions on the other sides are elliptic, too. If instead of $p$ we take another line $x$ (not through a vertex), we obtain other pairs of these involutions; hence the points $X_a$, $X_b$, $X_c$ (determined on the sides by the pole $X$) lie in the respective segments $BC/A_x$, $CA/B_x$, $AB/C_x$. Thus $X$ lies in the region $ABC/x$ and cannot lie on $x$. This means that there are no self-conjugate points, except possibly on a side of triangle $ABC$. But a self-

* von Staudt (Ref. 40, p. 133, §237); Enriques (Ref. 11, pp. 187–191).

conjugate point on a side would be an invariant point of the involution of conjugate points on that side, and we have just seen that such an involution is elliptic. This completes the proof.

**6·22** *If P does not lie in ABC/p, but in one of the other three regions, the polarity (ABC)(Pp) is hyperbolic.*

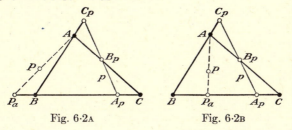

Fig. 6·2A　　　　　Fig. 6·2B

*Proof:* For definiteness, suppose that $P$ lies in the region penetrated by $p$ through sides $a$ and $b$, as in Fig. 6·2B. Then $A_p$ and $P_a$ both lie in the same segment $BC$, and thus the involution $(BC)(A_pP_a)$ is hyperbolic. Since this is the involution of conjugate points on $a$, its invariant points are self-conjugate.

Similarly, there are two self-conjugate points on $b$, though none on $c$.

**6·23** Corollary: *Both elliptic and hyperbolic polarities exist.*

The following temporary definitions will be found helpful in Sec. 6·3 but thereafter will be superseded: Let a point that is not self-conjugate be called an $E$ point or an $H$ point according as the involution of conjugate lines through it is elliptic or hyperbolic, and let a line that is not self-conjugate be called an $e$ line or an $h$ line according to the nature of the involution of conjugate points on it. We see at once that in an *elliptic* polarity every point is an $E$ point and every line is an $e$ line; and it emerges from the above proof of 6·22 that any self-polar triangle

Elliptic polarity　　Hyperbolic polarity

Fig. 6·2c

for a *hyperbolic* polarity has two $h$ sides and one $e$ side and consequently two $H$ vertices and one $E$ vertex. (Of course, the pole of an $e$ line or $h$ line is an $E$ point or $H$ point, respectively.) Self-polar triangles in the two cases are represented diagrammatically in Fig. 6·2c.

### EXERCISES

1. Carry out the construction 5·64 in the special case when $P$ lies on $p$. Observe that $X$ lies outside the region $ABC/x$.

2. Show that the two self-conjugate points on an *h* side (of a self-polar triangle for a hyperbolic polarity) lie respectively on the two self-conjugate lines through the opposite *H* vertex.

3. Which are the *E* vertex and *e* side of the triangle *ABC* in Fig. 5·6B?

### 6·3 How a Hyperbolic Polarity Determines a Conic.

Any line that is not self-conjugate may be used as a side of a self-polar triangle. Thus, in the case of a hyperbolic polarity, every point on an *e* line is an *H* point, but an *h* line contains points of both types, separated by the two self-conjugate points (which are the invariant points of the involution of conjugate points on the *h* line). Dually, the pencil of lines through an *E* point consists entirely of *h* lines, but the pencil of lines through an *H* point contains both types, separated by the two self-conjugate lines. Thus we have through an *E* point a pencil of *h* lines, each containing two self-conjugate points, and on an *e* line a range of *H* points, through each of which two self-conjugate lines can be drawn. Hence:

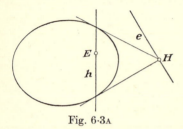

Fig. 6·3A

**6·31** *A hyperbolic polarity admits an infinity of self-conjugate points and an infinity of self-conjugate lines.*

Following von Staudt, we define a *conic* to be the locus of self-conjugate points in a hyperbolic polarity.* An *h* line (meeting the conic in two points, as in Fig. 6·3A) is a *secant*. A self-conjugate line (meeting the conic only at its pole) is a *tangent*, and the pole is its *point of contact*. An *e* line (containing no self-conjugate points) is an *exterior line*. The pole of a secant is an *exterior point* (*H* point), which lies on two tangents (the invariant lines of the involution of conjugate lines through it). Finally, the pole of an exterior line is an *interior point* (*E* point), which is characterized by the fact that no tangents can be drawn through it.

Thus a conic is essentially a self-dual figure: it is the locus of self-conjugate points and also the envelope of self-conjugate lines. Any of its properties can immediately be dualized by applying the polarity that defines it.

By our remarks about *H* points and *e* lines, every point on an exterior line is an exterior point. Dually, every line through an interior point is a secant. Again, any point on a tangent, except the point of contact, is exterior; and every line through a point on the conic, except the tangent there, is a secant.

* von Staudt (Ref. 40, p. 137, §246); Enriques (Ref. 11, pp. 199–201, 261–262).

## EXERCISES

1. Show that every point on a tangent is conjugate to the point of contact. Dually, the tangent itself is conjugate to any line through the point of contact.

2. Show that the polar of any exterior point joins the points of contact of the two tangents that can be drawn through the point. Dually, the pole of a secant $PQ$ is the point of intersection of the tangents $p$ and $q$.

3. If $PQR$ is a triangle inscribed in a conic, the tangents at $P$, $Q$, $R$ form a triangle circumscribed about the conic. Prove that these are a pair of Desargues triangles. (*Hint:* Use 5·61.)

4. If the tangents to a given conic meet a second conic in pairs of points, show that the tangents to the second conic at these pairs of points meet on a third conic. (*Hint:* Any construction for the first conic will be transformed, by the polarity wo the second, into the dual construction for the third.)

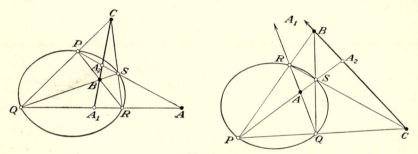

Fig. 6·4A

## 6·4 Conjugate Points and Conjugate Lines.* The following "harmonic property" can be traced back to Apollonius:

**6·41** *Any two conjugate points on a secant $PQ$ are harmonic conjugates wo $P$ and $Q$.*

*Proof:* The self-conjugate points $P$ and $Q$ are the invariant points of the involution of conjugate points on the line $PQ$ (see 4·63).

Dually,

**6·42** *Any two conjugate lines through an exterior point are harmonic conjugates wo the tangents that can be drawn through the point.*

The following theorem will enable us to construct the polar of a given point wo a given conic:

**6·43** *If a quadrangle is inscribed in a conic, its diagonal triangle is self-polar.*

*Proof:* Let the diagonal points of the inscribed quadrangle $PQRS$ be

$$A = PS \cdot QR, \qquad B = QS \cdot RP, \qquad C = RS \cdot PQ$$

* von Staudt (Ref. 40, pp.139–140, §§249–251).

as in Fig. 6·4A. The line $BC$ meets the sides $QR$ and $PS$ in points $A_1$ and $A_2$ such that $H(QR, AA_1)$ and $H(PS, AA_2)$. By 6·41, $A_1$ and $A_2$ are conjugate to $A$. Hence the line $BC$, which joins them, is the polar of $A$. Similarly $CA$ is the polar of $B$ and $AB$ of $C$.

Hence:

**6·44** *To construct the polar of a given point $A$, not on the conic, draw any two secants $QR$ and $PS$ through $A$; then the polar is*

$$(QS \cdot RP)(RS \cdot PQ)$$

In other words, we draw two secants through $A$ to form an inscribed quadrangle with diagonal triangle $ABC$, and then the polar of $A$ is $BC$.

The dual construction presupposes that we know the tangents from any given exterior point. This presents no serious difficulty (since

Fig. 6·4B

their points of contact lie on the polar of the given point); but the tangents are not immediately apparent, for the simple reason that we are in the habit of dealing with loci rather than envelopes. If we insist on regarding the conic as a locus, we may construct the pole of a given line as the point of intersection of the polars of any two points on the line. Then:

**6·45** *To construct the tangent at a given point $A$ on the conic, join $A$ to the pole of an arbitrary line through $A$.*

Theorem 6·43 and its dual may be neatly brought together as follows:

**6·46** *A quadrangle $PQRS$, inscribed in a conic, has the same diagonal triangle as the quadrilateral of tangents $pqrs$.*

*Proof:* Defining $A$, $B$, $C$ as before, we observe that $A$ lies on the polars of both $q \cdot r$ and $p \cdot s$ (see Fig. 6·4B). Hence the diagonal $(q \cdot r)(p \cdot s)$ of the quadrilateral is the polar of $A$, that is, it coincides with $BC$. Similarly the other two diagonals are $CA$ and $AB$.

## EXERCISES

1. Let $B$ and $C$ be two conjugate points wo a given conic. Let an arbitrary line through $C$ meet the conic in $P$ and $Q$, while $BP$ and $BQ$ meet the conic again in $R$ and $S$, respectively. Prove that $C$, $R$, $S$ are collinear.

2. If $PQR$ is a triangle inscribed in a conic, show that infinitely many self-polar triangles can be found having one vertex on each side of $PQR$.

3. Show that infinitely many triangles can be inscribed in a given conic in such a way that each side passes through one vertex of a given self-polar triangle.

4. Prove that a conic is transformed into itself by any harmonic homology whose center is the pole of its axis.

5. If the six points $P$, $P'$, $Q$, $Q'$, $R$, $R'$ of the Desargues configuration (Fig. 2·2A) lie on a conic, prove that $O$ is the pole of the line $ABC$ wo that conic. Deduce that the homology of 5·24 is then harmonic, so that the polarities $(PQR)(Oo)$ and $(P'Q'R')(Oo)$ coincide. Show also that any line through $O$ meets the six sides of the two triangles in a quadrangular set of points.[*]

**6·5  Two Possible Definitions for a Conic.[†]**  We have followed von Staudt in defining a conic as the locus of self-conjugate points in a hyperbolic polarity. Another definition, often used, is Steiner's (1832): A conic is the locus of the point of intersection of corresponding lines of two projective (but not perspective) pencils. We proceed to reconcile these two definitions. Theorem 6·52 will show that every von Staudt conic is a Steiner conic, and 6·54 will show that every Steiner conic is a von Staudt conic.

As a first step we need:

**6·51**  Seydewitz's Theorem: *If a triangle is inscribed in a conic, any line conjugate to one side meets the other two sides in conjugate points.*

*Proof:* Consider an inscribed triangle $PQR$. Any line $c$ conjugate to $PQ$ is the polar of some point $C$ on $PQ$. Let $RC$ meet the conic again in $S$, as in Fig. 6·4A. By 6·43, $c$ joins the points

$$A = PS \cdot QR, \qquad B = QS \cdot RP$$

These conjugate points $A$ and $B$ are the intersections of $c$ with the sides $QR$ and $RP$ of the given triangle.

**6·52**  Steiner's Theorem: *Let lines $x$ and $y$ join a variable point on a conic to two fixed points on the same conic; then $x \barwedge y$.*

---

* Mathews (Ref. 25, p. 338, Exercise 104).

† von Staudt (Ref. 40, pp. 141–144, §§253–258).

*Proof:* The tangents $p$ and $q$, at the fixed points $P$ and $Q$, intersect in $D$, the pole of $PQ$. Let $c$ be a fixed line through $D$ (but not through $P$ or $Q$), meeting $x$ and $y$ in $B$ and $A$, as in Fig. 6·5A. By 6·51, $BA$ is a pair of the involution of conjugate points on $c$. Hence, when the point $x \cdot y$ varies on the conic,

$$x \;\overline{\wedge}\; B \;\overline{\wedge}\; A \;\overline{\wedge}\; y$$

We shall prove the converse theorem with the help of the following lemma:

**6·53**   *A conic is determined when three points on it and the tangents at two of these are given.*

Let $P$, $Q$, $R$ be the given points, $PD$ and $QD$ the given tangents, and $C_1$ the harmonic conjugate of $C = PQ \cdot RD$ wo $P$ and $Q$, as in Fig. 6·5B. Consider the definite correlation that transforms the four points $P$, $Q$, $R$, $D$ into the four lines $PD$, $QD$, $RC_1$, $PQ$ and consequently trans-

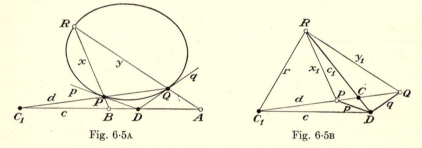

Fig. 6·5A                          Fig. 6·5B

forms $PQ$ into $D$, $RD$ into $C_1$, and $C$ into $C_1D$. This induces in $PQ$ a projectivity $PQC \;\overline{\wedge}\; PQC_1$. Since $H(PQ, CC_1)$, this is the involution $PQCC_1 \;\overline{\wedge}\; PQC_1C$; hence the correlation transforms $C_1$ into $CD$, and is a polarity (by 5·62 applied to triangle $DCC_1$). Since the polars of $P$, $Q$, $R$ are $PD$, $QD$, $RC_1$, the polarity determines a conic having the desired properties.

**6·54**   **Steiner's Construction:** *Let variable lines $x$ and $y$ pass through fixed points $P$ and $Q$ in such a way that $x \;\overline{\wedge}\; y$ but not $x \;\overline{\overline{\wedge}}\; y$; then the locus of $x \cdot y$ is a conic through $P$ and $Q$.*

*Proof:* Since the projectivity $x \;\overline{\wedge}\; y$ is not a perspectivity, the line $d = PQ$ does not correspond to itself. Hence there exist lines $p$ and $q$ such that the projectivity relates $p$ to $d$, and $d$ to $q$. By 6·53 we can find a conic touching $p$ at $P$, $q$ at $Q$ and passing through any other particular position $x_1 \cdot y_1$ of the variable point $x \cdot y$. By 6·52 the conic determines a projectivity between pencils through $P$ and $Q$; and this must coincide with the given projectivity $x \;\overline{\wedge}\; y$, since it relates the three lines $x_1$, $p$, $d$ through $P$ to the three lines $y_1$, $d$, $q$ through $Q$.

**6·55**  Corollary: *If a projectivity between lines through P and Q has the effect* $xpd \; \overline{\wedge} \; ydq$, *where d is PQ, then p and q are the tangents at P and Q to the locus of* $x \cdot y$.

**6·56**  *A unique conic can be drawn through five given points, provided that no three of the points are collinear.*

*Proof:* The two points $P$, $Q$ and three positions of $x \cdot y$ determine a projectivity $x_1 x_2 x_3 \; \overline{\wedge} \; y_1 y_2 y_3$, which yields a conic through the five points (by 6·54). Conversely, if a point on any conic through the five points is joined to $P$ and $Q$ by lines $x$ and $y$, we must have $xx_1 x_2 x_3 \; \overline{\wedge} \; yy_1 y_2 y_3$ (by 6·52), hence the conic is unique.

(Lemma 6·53 may be regarded as the special case that arises when two pairs of the five given points coincide.)

Fig. 6·5c

The duals of 6·51 and 6·54 are sufficiently important to be stated explicitly:

**6·57**  *If a triangle is circumscribed about a conic, any point conjugate to one vertex is joined to the other two vertices by conjugate lines.*

**6·58**  *Let points X and Y vary on fixed lines p and q in such a way that* $X \; \overline{\wedge} \; Y$ *but not* $X \; \overline{\overline{\wedge}} \; Y$; *then the envelope of the line XY is a conic touching p and q.* (This was proved by Chasles in 1828.)

It follows from 6·53 that infinitely many conics can be drawn to touch two fixed lines at two fixed points. Such conics are said to have *double contact* (with one another).

**6·59**  *Of the conics that touch two given lines at given points, those which meet a third line (not through either of the points) do so in pairs of an involution.*

*Proof:* Let such a conic touch $AB$ at $P$, $AC$ at $Q$ and meet $BC$ in $R$ and $S$, as in Fig. 6·5c. Let the polar of the point $M = PQ \cdot BC$ meet $PQ$ in $L$, and $BC$ in $N$. Then by 6·41 we have H($RS$, $MN$), H($PQ$, $ML$), and consequently H($BC$, $MN$). Hence $RS$ is a pair of the hyperbolic involution $(MM)(BC)$, whose invariant points are $M$ and $N$.

## EXERCISES

1. Dualize 6·52 and 6·53.

2. If variable points $X$ and $Y$ on fixed lines $p$ and $q$ are conjugate for a given polarity, while the point $p \cdot q$ is not self-conjugate, prove that the line $XY$ envelops a conic.[*]

3. Show that the four points $P$, $Q$, $R$, $S$ of 6·46 lie on a conic through $p \cdot s$ and $q \cdot r$.[†]

4. Let $A$ and $B$ be a pair of conjugate points wo a conic, and let $PQ$ be a secant conjugate to $AB$. Prove that $AQ$ and $BP$ meet on the conic. (*Hint: BP meets the conic again in the point $R$ of Sec. 6·4, Exercise 1.*)

5. Consider a variable conic inscribed in a given quadrilateral. Show that the line joining its points of contact with two sides of the quadrilateral passes through a fixed point. Identify this point.[‡]

6. Measure off points $X_0, X_1, \ldots, X_5$ at equal intervals along a line, and $Y_0, Y_1, \ldots, Y_5$ similarly along another line through $X_5 = Y_0$. The joins $X_n Y_n$ visibly envelop a conic, in accordance with 6·58. (Of course, this method of setting up a projectivity is not playing the game. We shall have to wait till Chap. 8 to see why it is valid, and why the conic is a parabola.)

**6·6  Construction for the Conic through Five Given Points.**  The following construction was given by Braikenridge in 1733, but his priority was contested by Maclaurin in a rather disagreeable controversy:

**6·61**  *If the sides of a variable triangle pass through three fixed non-collinear points, while two vertices run along fixed lines, the third vertex will trace a conic through two of the given points.*

*Proof:* Let $LMC'$ be the variable triangle whose sides $x = MC'$, $y = LC'$, $z = LM$ pass through fixed points $A$, $B$, $N$, while the vertices $L$ and $M$ run along fixed lines $CB'$ and $CA'$, as in Fig. 6·6A. Then

$$x \underset{\overline{A}}{\overset{CA'}{=}} z \underset{\overline{A}}{\overset{CB'}{=}} y$$

The projectivity $x \overline{\overline{\wedge}} y$ could be a perspectivity only if $N$ lay on $AB$. By 6·53 the locus of $C' = x \cdot y$ is a conic through $A$ and $B$.

This conic passes also through $A'$ (on $NB$), $B'$ (on $NA$), and $C$. For when $z$ coincides with $NB$, $y$ does the same, while $M$ and $C'$ coincide with $A'$. Similarly, when $z$ coincides with $NA$, $C'$ coincides with $B'$. Finally, when $z$ coincides with $NC$, the points $L$, $M$, $C'$ all coincide with $C$.

---

[*] Chasles (Ref. 4, pp. 10, 137).
[†] Chasles (Ref. 4, p. 138).
[‡] Graustein (Ref. 14, p. 325).

In other words, given five points $A$, $B$, $C$, $A'$, $B'$ of which no three are collinear, we may construct any number of positions of a sixth point $C'$ on the conic $ABCA'B'$ in the following manner: Through the point $N = AB' \cdot BA'$ draw an arbitrary line $z$, meeting $CB'$ in $L$, and $CA'$ in $M$. Then $C' = AM \cdot BL$ is another point on the conic.

Fig. 6·6A

Fig. 6·6B

Fig. 6·6c

Fig. 6·6D

By 6·55 the tangent at $A$ (Fig. 6·6A) is the position taken by $x$ when $y$ coincides with $BA$, so that $L$ is $AB \cdot CB'$, as in Fig. 6·6B. Hence:

**6·62**   *The tangent at $A$ to the conic $ABCA'B'$ is the line $AM$ where* $M = CA' \cdot (AB' \cdot BA')(AB \cdot CB')$.

### EXERCISES

1. Dualize 6·61, and deduce a construction for any number of tangents to the conic determined by five given lines, no three concurrent (see Fig. 6·6c).

2. Dualize 6·62 to obtain a construction for the point of contact of any one of five given tangents (see Fig. 6·6D).

3. Observe that Fig. 4·8A is the same as Fig. 6·6A if we name the points of the former as follows:

$$A = a \cdot b, \quad B = a' \cdot b', \quad C = c \cdot c', \quad A' = a \cdot a', \quad B' = b \cdot b', \quad C' = x \cdot x'$$
$$L = s \cdot x', \quad M = r \cdot x, \quad N = a' \cdot b$$

4. Prove that, if the three pairs of opposite sides of a hexagon meet in three collinear points, then the six vertices lie on a conic. (*Hint:* Consider the hexagon $AB'CA'BC'$ of Fig. 6·6A.) Dually, if the three diagonals of a hexagon are concurrent, the six sides touch a conic.

5. Given a pair of Desargues triangles, show that the six points in which the sides of one triangle meet the noncorresponding sides of the other lie on a conic. Dually, the six lines joining the vertices of one triangle to the noncorresponding vertices of the other are tangents of a conic.

**6·7 Two Triangles Inscribed in a Conic.***  The following results are interesting in themselves, besides illustrating the use of Steiner's construction.

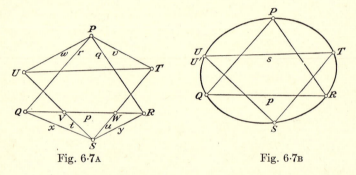

Fig. 6·7A          Fig. 6·7B

**6·71**  *If two triangles are self-polar for a given polarity, their six vertices lie on a conic and their six sides touch another conic.*

*Proof:* The given self-polar triangles $PQR$ and $STU$ determine further points and lines

$$V = p \cdot t, \quad W = p \cdot u, \quad v = PT, \quad w = PU, \quad x = QS, \quad y = RS$$

as in Fig. 6·7A. With the help of 2·71, we find

$$rqvw \ \overline{\wedge}\ RQVW \ \overline{\wedge}\ QRWV \ \overline{\wedge}\ xyut$$

Therefore, by 6·54, the four points $Q$, $R$, $T$, $U$ lie on a conic through $P$ and $S$. The last part of the theorem follows by duality (or may be proved similarly, using 6·58).

**6·72**  *Any two triangles inscribed in a conic are self-polar for some polarity.*

* von Staudt (Ref. 40, pp. 174–175, §§299, 300).

*Proof:* Let $PQR$ and $STU$ be the two triangles. Consider the polarity $(PQR)(Ss)$, where $s$ is $TU$. Let the polar of $T$ meet $s$ in $U'$, as in Fig. 6·7B. Then $PQR$ and $STU'$ are two self-polar triangles. By 6·71, $U'$ lies on the conic $PQRST$. (It also lies on $s$.) But so does $U$. Hence $U'$ coincides with $U$.

**6·73** Desargues' Involution Theorem: *Of the conics that can be drawn through the vertices of a given quadrangle, those which meet a given line (not through a vertex) do so in pairs of the quadrangular involution.*

*Proof:* By 6·72 the points $T$ and $U$ in which any conic through $P$, $Q$, $R$, $S$ meets the given line $s$ are a pair of the involution of conjugate points on $s$ for the polarity $(PQR)(Ss)$. Since $PS$ meets $s$ in a point conjugate to $p \cdot s$, the pairs of opposite sides of the quadrangle meet $s$ in pairs of this same involution (cf. 6·59).

Conversely, any two points on $s$ that are paired in the quadrangular involution lie on a conic through $P$, $Q$, $R$, $S$, provided that we include as *degenerate conics* the line pairs such as $PS$ and $QR$. This idea of regarding two lines as forming a degenerate conic enables us to delete the clause *but not $x \overline{\overline{\wedge}} y$* from 6·54, which may now be expressed as follows: *A conic is the locus of a point that is joined to two fixed points by corresponding lines of two projectively related pencils.* For although when $x \overline{\overline{\wedge}} y$ the locus of $x \cdot y$ is strictly only one line, we must admit that any point collinear with $A$ and $B$ shares with the points of that locus the property of being joined to $A$ and $B$ by corresponding lines of the related pencils.

### EXERCISES

1. Observe how 6·71 and 6·72 serve to establish the following theorem, due to Steiner: *If two triangles are inscribed in a conic, their six sides touch a conic (and conversely).*

2. Prove that a given line touches at most two of the conics through $P$, $Q$, $R$, $S$.

3. The formal self-duality of the Desargues configuration (Fig. 2·2A and B) continues to hold if we interchange the names of $P$ and $P'$, $Q$ and $Q'$, $R$ and $R'$. Prove that this *duality* cannot be realized as a *polarity* (cf. 5·71) unless these six points lie on a conic (as in Sec. 6·4, Exercise 5).

**6·8  Pencils of Conics.**\* We saw in Sec. 5·8 that the polarities $(ABC)(Pp)$, where $p$ turns about a fixed point $P'$, have the property that the polars of an arbitrary point $X$ all pass through a corresponding point $X'$. In other words, the pencil of polarities determines a corre-

---

\* This section may be omitted on first reading or in a short course.

spondence $X \to X'$ between points that are simultaneously conjugate for all the polarities. Now, if $X$ runs along a fixed line $o$, how does $X'$ behave? The answer will emerge while we are proving the following theorem:

**6·81**    *The locus of poles of a fixed line wo a pencil of polarities is a conic or a line.*

*Proof:* By the dual of 5·81 the locus is a line if the pencil is a "self-dual system"; but we are more interested in the general case. Let $O_1$ and $O_2$ be the poles of the given line $o$ wo two particular polarities in the pencil (see Fig. 6·8A). When the point $X$ varies on $o$, its two polars

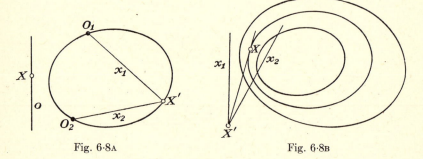

Fig. 6·8A                              Fig. 6·8B

$x_1 = O_1X'$ and $x_2 = O_2X'$ rotate about the fixed points $O_1$ and $O_2$ and we have

$$x_1 \barwedge X \barwedge x_2$$

Three particular positions for $X$ are where $o$ cuts the sides of the common self-polar triangle $ABC$ of the polarities; then the conjugate points $X'$ are the vertices of that triangle, in turn. Hence the locus of $X'$ (while $X$ runs along $o$) is the conic $ABCO_1O_2$, which is nondegenerate provided that no three of these five points are collinear. Each position of $X'$ depends only on $X$ and the pencil, not on the two selected polarities. Hence any other polarity in the pencil will give a pole $O$ on this same conic. In other words, the conic is both the locus of $X'$ (when $X$ varies on $o$) and the locus of the pole $O$ (when the polarity varies in the pencil).

The possible cases of degeneracy remain to be investigated. We shall find that, when the conic degenerates into two lines, the loci of $X'$ and $O$ no longer coincide but $X'$ runs along one line and $O$ along the other.

One possibility is that $O_1$ might lie on a side of triangle $ABC$, say $a$. This can happen only if the given line $o$ passes through $A$, in which case not only $O_1$ but also $O_2$ lies on $a$. The position $A$ for $X$ makes $x_1$

and $x_2$ coincide with $a$; therefore $x_1 \overline{\overline{\wedge}} x_2$, and the locus of $X'$ is a line through $A$. But the locus of $O$ is the line $a$.

On the other hand, if $O_1$ and $O_2$ are collinear with a vertex, say $A$, (but do not lie on a side), we have $AO_1 = AO_2$, so that the point $a \cdot o$ has the same polar for both polarities. This is the case of a self-dual system of polarities. The locus of $X'$ is the side $a$, while that of $O$ is a certain line through $A$. Here we have assumed that $o$ does not pass through the special vertex $A$; but if it does, its pole $O$ is the same for all the polarities and $X'$ constantly coincides with $O$.

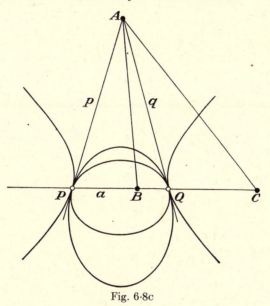

Fig. 6·8c

**6·82** *Every pencil of polarities determines a pencil of conics: one conic through each point of general position.*

*Proof:* Consider once more the polarities $(ABC)(Pp)$, with $p$ through the fixed point $P'$. Since one possible position for $p$ is $P'P$, one of the polarities determines a conic that touches $PP'$ at $P$. In fact, every point $X$ (not on a side of triangle $ABC$) lies on such a conic (touching $XX'$ at $X$).

In particular, a self-dual system of polarities (see 5·82) determines a *self-dual system of conics* (*e.g.*, Fig. 6·8B or c).

It may happen that two conics of a pencil have a common point. Then this point is self-conjugate for two, and therefore all, of the polarities.

If such a common point $P$ lies on the side $a$ of the common self-polar triangle $ABC$, then all the conics have the same tangent $PA$ at $P$ and the same tangent $QA$ at $Q$, the harmonic conjugate of $P$ wo $B$ and $C$ (see Fig. 6·8c). Thus, *if two conics of a self-dual system have a common point, all of them have double contact.*

On the other hand, if the common point $P$ does not lie on a side of triangle $ABC$, then we can find (as in 2·42) a definite quadrangle $PQRS$ having $ABC$ for its diagonal triangle. All four vertices of this quadrangle are self-conjugate; for example, $Q$ (being the harmonic conjugate of $P$ wo the two conjugate points $C$ and $C_1$ of Fig. 2·4c) is the second invariant point of the involution of conjugate points on $CP$. Hence in this case the pencil consists of *all the conics circumscribed to a quadrangle* (as in 6·73), and we call it a *quadrangular* pencil of conics.

The "double-contact" system may be regarded as a limiting case of this quadrangular pencil, when the points $P$, $Q$, $R$, $S$ approach coincidence in pairs, $PS$ and $QR$, in such a way that the lines $PS$ and $QR$ have definite limiting positions $p$ and $q$. For then the conics will all touch $p$ at $P$, and $q$ at $Q$.

As a special case of 6·81, we have the following:

**6·83**  *Given a quadrangular pencil of conics and a line not through a diagonal point of the quadrangle, the locus of poles of the line is a conic.*

This conic, which is not only the locus of poles $O$ of the given line $o$ but also the locus of the point $X'$, which is conjugate to a variable point $X$ on $o$ wo all the conics, is called the *nine-point conic* of the quadrangle wo the line $o$. For it contains nine special points: the three diagonal points $A$, $B$, $C$ and one further point on each side; *e.g.*, on $PQ$ the harmonic conjugate of $o \cdot PQ$ wo $P$ and $Q$. These nine points are the positions taken by $X'$ when $X$ lies, in turn, on the sides of the diagonal triangle and of the quadrangle itself.

Since conics are ultimately defined in terms of incidence, any collineation will transform a conic into a conic. In particular, a homology whose center and axis are the $A$ and $a$ of 5·82 leaves the self-dual system invariant: if it is a harmonic homology, it leaves each conic invariant, but otherwise it transforms each conic into another. If the common involution is elliptic, $A$ is interior to all the conics and each line through $A$ meets them all. Hence in this case, by 5·21, any two conics of the system are related by such a homology. But if the involution is hyperbolic, so that $A$ is outside all the conics, then any two lines through $A$, separated by $p$ and $q$, will each be a secant for some of the conics and an exterior line for the rest. Thus a system of conics having double contact falls into two subsystems, such that any two conics of the same subsystem are related by a homology with center $A$ and axis $a$. (These subsystems are represented by ellipses and hyperbolas in Fig. 6·8c, but of course that distinction belongs to affine geometry; in projective geometry they are exactly alike.) On the other hand, each line through $P$ meets all the conics (of both subsystems); hence any two conics of the whole system are related by a homology with center $P$ and axis $q$ (and likewise by one with center $Q$ and axis $p$). In either case:

**6·84**   *Any two conics of a self-dual system are related by a homology.*

## EXERCISES

1. Theorem 6·82 shows that every pencil of polarities includes infinitely many hyperbolic polarities. Prove that the pencil consists entirely of hyperbolic polarities if a common pair of conjugate points ($P$ and $P'$) both lie in the same one of the four regions determined by the common self-polar triangle $ABC$; but infinitely many elliptic polarities occur as well if such conjugate points lie in different regions. (In the case of a self-dual system the same distinction depends on whether the common involution is hyperbolic or elliptic.)

2. In the presence of a general pencil of polarities any line $o$ determines a conic (the locus of poles of $o$; see 6·81). Show that a pencil of lines $o$ determines a pencil of such conics.

3. If two quadrangles have the same diagonal points, prove that either they share a pair of opposite sides or their eight vertices lie on a conic. (*Hint:* Consider the conic determined by one quadrangle and one vertex of the other.)

4. Using 6·46 and the dual of Exercise 3, prove that, if two conics intersect in four points, the eight tangents at these points either pass by fours through two points or touch a conic (*Salmon's conic*).

5. For the nine-point conic show that three pairs of the nine points (*e.g.*, the harmonic conjugate of $o·PQ$ on $PQ$ and that of $o·RS$ on $RS$) are joined by lines conjugate to $o$.

# CHAPTER 7

# PROJECTIVITIES ON A CONIC

This chapter deals with those properties of a conic which may be most readily derived by means of the notion that the points on the conic form a "range," resembling in many ways the points on a line. Pascal's theorem is the most famous instance; but its original proof must have been different. The idea of projectivity on a conic is due to Bellavitis (1838). We shall see that the construction for such a projectivity is simpler than for a projectivity on a line. In fact, some authors, such as Holgate, rearrange the material so as to treat ranges on a conic before ranges on a line. Involutions are especially easy to deal with, for the joins of pairs of corresponding points are concurrent, as we shall see in Sec. 7·5.

**7·1 Generalized Perspectivity.** Steiner's theorem (our 6·52) enables us to regard the points on a conic as a *range* that can be related to an ordinary range or pencil or to another range on the same (or another) conic. Thus, if variable points $R$ and $R'$ on a conic are joined to a fixed point *on the same conic* by lines $x$ and $x'$, where $x \ \overline{\wedge}\ x'$, then we are justified in writing

$$R \ \overline{\wedge}\ R'$$

since the fixed point $P$ could just as well be replaced by another such point $Q$ (on the same conic), as in Fig. 7·1A.

If $x$ meets a fixed line (not through $P$) in $B$, we write

$$R \ \overset{P}{\overline{\wedge}}\ B$$

and call this relation a *generalized perspectivity*. (Always remember that $P$ has to lie on the conic.) When $R$ coincides with $P$, $x$ is the tangent there (by 6·55) and $B$ is where this tangent meets the fixed line.

This notion makes it easy to prove the following theorem:

**7·11** *If the six vertices of two triangles lie on a conic, the six sides touch a conic (and conversely).*

*Proof:* The given inscribed triangles $PQR$ and $STU$ determine further points

$$V = QR \cdot SU, \qquad W = QR \cdot ST, \qquad X = PR \cdot TU, \qquad Y = PQ \cdot TU$$

(see Fig. 6·7A, where the points $X$ and $Y$ should be marked), and we have

$$RQVW \overset{S}{\barwedge} RQUT \overset{P}{\barwedge} XYUT$$

Since $P$ and $S$ are distinct, the projectivity $RQVW \barwedge XYUT$ is not a perspectivity. Therefore, by 6·58, the lines $RX$, $QY$, $VU$, $WT$ are tangents of a conic that also touches $RQ$ and $UT$. (The converse is the dual theorem.) (For another proof, see Sec. 6·7, Exercise 1.)

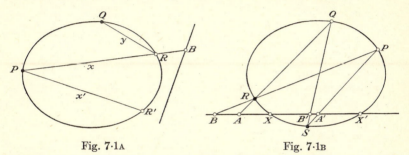

Fig. 7·1A             Fig. 7·1B

As another application of this method, here is an alternative proof* of Desargues' involution theorem (our 6·73). Let the given line meet the sides $QR$, $PS$, $RP$, $QS$ of the quadrangle, and one of the conics, in the points $A$, $A'$, $B$, $B'$, $X$ and $X'$, as in Fig. 7·1B. Then

$$ABXX' \overset{R}{\barwedge} QPXX' \overset{S}{\barwedge} B'A'XX'$$

Hence, by 4·67 and 4·71, $XX'$ is a pair of the quadrangular involution $(AA')(BB')$.

This proof has the advantage of remaining valid when $P$ coincides with $S$ (or $Q$ with $R$) so that the conics form a "contact" pencil instead of a quadrangular pencil. It even remains valid when $P = S$ and $Q = R$, so that the conics form a "double-contact" system, as in 6·59. In each case, those conics of the pencil which meet a line of general position do so in pairs of an involution.

### EXERCISE

Establish the relation $PQRS \barwedge QPSR$ for any four points on a conic. (*Hint:* Use 2·71.)

---

* von Staudt (Ref. 40, p. 176, §301).

**7·2   Pascal and Brianchon.**   At the age of sixteen Pascal wrote an extensive treatise on conics, which, unhappily, is now lost. Leibniz, who saw it, says that it included the famous theorem of the *hexagramma mysticum*, which is the converse of Braikenridge's construction (our 6·61). We do not know how Pascal proved it; but various proofs have been devised since his time.

**7·21   Pascal's Theorem**: *If a hexagon is inscribed in a conic, the three pairs of opposite sides meet in three collinear points.*

*Proof:** Let $AB'CA'BC'$ be the hexagon, so that the points to be proved collinear are

$$L = BC' \cdot CB', \qquad M = CA' \cdot AC', \qquad N = AB' \cdot BA'$$

Fig. 7·2A             Fig. 7·2B

as in Fig. 7·2A. Using further points $J = AC' \cdot BA'$ and $K = BC' \cdot CA'$, we have

$$A'NJB \overset{A}{\barwedge} A'B'C'B \overset{C}{\barwedge} KLC'B$$

Thus $B$ is an invariant point of the projectivity $A'NJ \barwedge KLC'$, which accordingly is a perspectivity, namely, $A'NJ \overset{M}{\barwedge} KLC'$ (since the joins $A'K$ and $JC'$ pass through $M$). Hence $NL$ passes through $M$.

Note the close analogy with Pappus's theorem (our 4·31), which is Pascal's theorem as applied to a degenerate conic. (Some German authors call Pappus's theorem "Pascal's theorem.") Anyone who dislikes the generalized perspectivity may deduce

$$A'NJB \barwedge KLC'B$$

from the observation that the lines $AA'$, $AB'$, $AC'$, $AB$ are projectively related to the respective lines $CA'$, $CB'$, $CC'$, $CB$.

* von Staudt (Ref. 40, p. 143, §257); Enriques (Ref. 11, p. 225).

The dual of Pascal's theorem was discovered by Brianchon, nearly 170 years later.

**7·22**  Brianchon's Theorem: *If a hexagon is circumscribed about a conic, the three diagonals are concurrent* (see Fig. 7·2B).

The following is typical of many applications of Pascal's theorem:

**7·23**  *If $A$, $B$, $C$, $A'$, $B'$, $C'$ are six points on a conic, while the tangents at these points are $a$, $b$, $c$, $a'$, $b'$, $c'$, then the three lines $(a \cdot a')(BC \cdot B'C')$, $(b \cdot b')(CA \cdot C'A')$, $(c \cdot c')(AB \cdot A'B')$ are concurrent.*

*Proof:* Let the various points be named

$$A_1 = a \cdot a', \qquad B_1 = b \cdot b', \qquad C_1 = c \cdot c'$$
$$L = BC'' \cdot CB', \qquad M = CA' \cdot AC', \qquad N = AB' \cdot BA'$$
$$P = BC \cdot B'C', \qquad Q = CA \cdot C'A', \qquad R = AB \cdot A'B'$$

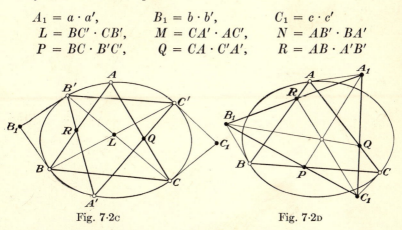

Fig. 7·2C                                    Fig. 7·2D

as in Fig. 7·2C. From the Pascal hexagon $CABC'A'B'$, the three points $L$, $Q$, $R$ are collinear. By 6·43, $L$ lies on the polar of $BB' \cdot CC'$, which is $B_1C_1$. Thus $L$ lies on both $B_1C_1$ and $QR$. Similarly $M$ lies on $C_1A_1$ and $RP$, $N$ on $A_1B_1$ and $PQ$. But $L$, $M$, $N$ are collinear (from the Pascal hexagon $AB'CA'BC'$). Hence $A_1B_1C_1$ and $PQR$ are Desargues triangles.

### EXERCISES

1. Verify that $A'RSBXX'$ is a Brianchon hexagon in Fig. 4·2A.

2. Show how the pentagon $AB'CA'B$ of 6·62 may be regarded as a limiting case of the Pascal hexagon when $C'$ coincides with $A$.

3. Show how the quadrangle $PQSR$ of 6·46 may be regarded as a limiting case of the Pascal hexagon when two vertices coincide at $Q$ and two others at $R$, so that the "hexagon" is $PQQSRR$. Similarly, the quadrilateral $prqs$ may be regarded as a limiting Brianchon hexagon $pprqqs$.

4. Let a conic be defined by four points and the tangent at one of them. Construct the tangent at another one of the four points.

5. Show how the triangle $PQR$ of Sec. 6·3, Exercise 3, may be regarded as a Pascal hexagon $PPQQRR$ or as a Brianchon hexagon $ppqqrr$.

6. Let $ABC$ be a triangle inscribed in a conic. Choosing any points $A_1$, $B_1$, $C_1$ on the tangents at $A$, $B$, $C$, let the sides of triangle $ABC$ meet the corresponding sides of triangle $A_1B_1C_1$ in points $P$, $Q$, $R$, as in Fig. 7·2D. Prove that the three lines $A_1P$, $B_1Q$, $C_1R$ are concurrent.*

**7·3   Construction for a Projectivity on a Conic.†**   In virtue of Steiner's theorem the whole theory of projectivities on a line can be carried over to projectivities on a conic; *e.g.*, such a projectivity may be *elliptic* (having no invariant point), *parabolic* (having one invariant point), or *hyperbolic* (having two invariant points), and in the last case it is either *direct* or *opposite* (according as it preserves or reverses the

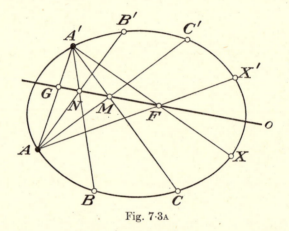

Fig. 7·3A

sense around the conic). The construction for the transform of a given point is actually easier than in Sec. 4·4, for now we can make use of an auxiliary line called the axis of the projectivity.

The *axis* of the projectivity $ABC \barwedge A'B'C'$ on a conic is the Pascal line of the hexagon $AB'CA'BC'$, namely, the line $o$ determined by any two of the three points $BC' \cdot CB'$, $CA' \cdot AC'$, $AB' \cdot BA'$.

**7·31**   *Given seven points $A$, $B$, $C$, $X$, $A'$, $B'$, $C'$ on a conic, we can locate the point $X'$ such that $ABCX \barwedge A'B'C'X'$ as the second intersection of the conic with the line $AF$, where $F$ is $XA' \cdot o$.*

*Proof:* If $AA'$ meets the axis $o$ in $G$ (Fig. 7·3A), we have

$$ABCX \overset{A'}{\barwedge} GNMF \overset{A}{\barwedge} A'B'C'X'$$

---

* This theorem, due to Kenneth Leisenring (a graduate student at the University of Michigan), was put into the more manageable form 7·23 by Alex Rosenberg.

† Enriques (Ref. 11, p. 251).

Thus the "cross joins" of any two pairs of corresponding points meet on the axis.

The general projectivity between ranges on two distinct lines (Fig. 4·3c) may be regarded as a degenerate case, the Pascal line of the conic becoming the Pappus line of the line pair.

**7·32**  *A projectivity on a conic is determined when its axis and one pair of corresponding points are given.*

*Proof:* Given the axis and $AA'$, we have $X \overset{A'}{\barwedge} F \overset{A}{\barwedge} X'$. In other words, for any $X$ on the conic, there is a unique $X'$ such that $AX' \cdot XA'$ lies on the axis.

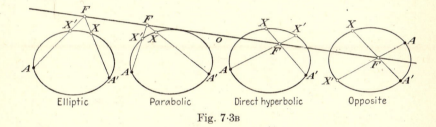

Fig. 7·3B

The only way in which $X'$ can coincide with $X$ is by $X$ lying on the axis. Hence:

**7·33**  *The invariant points (if any) of a projectivity on a conic are the common points of the axis and the conic.*

**7·34**  *The projectivity is elliptic, parabolic, or hyperbolic, according as its axis is an exterior line, a tangent, or a secant.*

A secant decomposes the conic into two arcs. Hence, by 4·42, a hyperbolic projectivity is direct or opposite according as two corresponding points lie both on the same arc or one on each. The four types of projectivity are illustrated in Fig. 7·3B.

**7·35**  *Any projectivity on a conic determines a collineation of the whole plane.*

*Proof:* By 6·53 a unique conic can be drawn to touch two given lines $a$ and $b$ at two given points $A$ and $B$ and to pass through another given point $C$. Consider the quadrangle $ABCD$, where $D$ is $a \cdot b$, and another such quadrangle $A'B'C'D'$ (determining the same conic). These two quadrangles are related by a unique collineation (see 5·12), which preserves the conic and induces the projectivity $ABC \overset{}{\barwedge} A'B'C'$ on it.

EXERCISE

Show that an involution on a conic determines a harmonic homology of the whole plane.

**7·4  Construction for the Invariant Points of a Given Hyperbolic Projectivity.**  Given a hyperbolic projectivity on a conic, we can easily locate its invariant points by drawing its axis (see 7·33). This

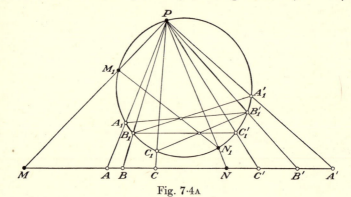

Fig. 7·4A

suggests the following construction (due to Steiner) for the invariant points of a given hyperbolic projectivity $ABC \overline{\wedge} A'B'C'$ on a line:

Draw any conic (in practice, a circle*), and project the given points from any point $P$ on the conic into $A_1$, $B_1$, $C_1$, $A_1'$, $B_1'$, $C_1'$ on the conic, as in Fig. 7·4A. Draw the axis $(A_1B_1' \cdot B_1A_1')(B_1C_1' \cdot C_1B_1')$, to meet the conic in $M_1$ and $N_1$. Project these points from $P$, back onto the original line, and we obtain the desired invariant points $M$ and $N$.

$$\text{Proof: } MNABC \overset{P}{\overline{\wedge}} M_1N_1A_1B_1C_1 \overset{}{\overline{\wedge}} M_1N_1A_1'B_1'C_1' \overset{P}{\overline{\wedge}} MNA'B'C'$$

EXERCISE

What will happen if we try to carry out this construction when the given projectivity is not hyperbolic?

**7·5  Involution on a Conic.†**  The involution $(AA')(BB')$ may be regarded as the special case of the projectivity $ABC \overline{\wedge} A'B'C'$ that arises when $C = B'$ and $C' = B$, as in Fig. 7·5A. The axis

$$o = (AB' \cdot BA')(AC' \cdot CA') = (AB' \cdot BA')(AB \cdot A'B')$$

---

\* The familiar process of reciprocation wo a circle is an instance of a polarity; therefore a circle is a conic. We shall return to this subject in Sec. 9·2.

† Cremona (Ref. 8, p. 160). For an alternative treatment see Enriques (Ref. 11, pp. 256–259) or Veblen and Young (Ref. 42, p. 222).

is one side of the diagonal triangle of the quadrangle $AA'BB'$, and its pole is the opposite vertex $O = AA' \cdot BB'$ of that triangle. Hence:

**7·51**    *The pairs of an involution of points on a conic are joined by concurrent lines; i.e., they are cut out by a pencil of secants.*

Conversely, any pencil of secants determines an involution, namely, $(AA')(BB')$, where $AA'$ and $BB'$ are any two of the secants.

The axis, being the polar of $O$, contains the point of intersection of the tangents at any two corresponding points. This could also be inferred by considerations of continuity; for the tangents at $A$ and $A'$

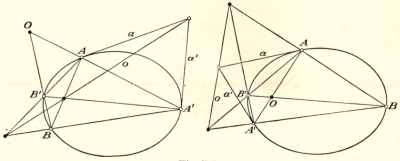

Fig. 7·5A

are the limiting positions of the cross joins of the pairs $AA'$ and $B'B$ when $B$ approaches $A$.

The point $O = AA' \cdot BB'$ is called the *center* of the involution $(AA')(BB')$. Since it is the pole of the axis, 7·34 implies the following:

**7·52**    *The center O of an involution is interior or exterior according as the involution is elliptic or hyperbolic, and in the latter case the invariant points are the points of contact of the two tangents that can be drawn through O.*

**7·53**    Corollary: *Four points on a conic satisfy $AA'//BB'$ if and only if the point $AA' \cdot BB'$ is interior.*

By 4·63 we now have:

**7·54**    *Two secants $AA'$ and $MN$ are conjugate lines if and only if $H(AA', MN)$.*

The following theorem will be used in Sec. 9·3, where we consider the axes of a conic in Euclidean geometry:

**7·55**    *Two involutions, one or both elliptic, on the same line, always have a common pair of corresponding points.*

*Proof:* Transfer the two involutions to a conic, by the method of Sec. 7·4, and let their centers be $O_i$ and $O_j$, as in Fig. 7·5B. Since at least one of these points is interior, their join $O_iO_j$ is a secant, meeting the conic in $M$ and $N$, say. Then $MN$ is a common pair of the two involutions on the conic (cf. Sec. 4·6, Exercise 4).

Incidentally, this shows that the product of the two involutions is hyperbolic, having $M$ and $N$ for invariant points. More generally, the product of *any* two involutions on a conic has for axis the join of the

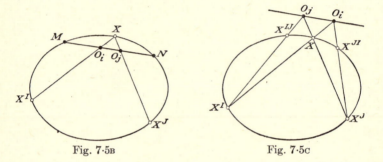

Fig. 7·5B                    Fig. 7·5c

two centers. For if the involutions transform $X$ into $X^I$ and $X^J$, respectively, as in Fig. 7·5c, their product

$$XX^IX^{JI} \barwedge X^{IJ}X^JX$$

has axis $(X^IX \cdot X^{JI}X^J)(XX^J \cdot X^IX^{IJ}) = O_iO_j$.

*EXERCISES*

1. What happens to the involution with center $O$ when the conic degenerates into a line pair? Where then is the axis of the involution?

2. Use 7·53 to obtain a quick answer to Exercise 3 of Sec. 4·7.

3. Given five points $A$, $A'$, $B$, $B'$, $C$, no three collinear, devise a linear construction for the point $C'$ on the conic $AA'BB'C$ that is paired with $C$ in the involution $(AA')(BB')$.

4. Adapt the method of Sec. 7·4 so as to construct the points $M$ and $N$ of 3·64. These are, of course, the invariant points of the hyperbolic involution $(AB)(CD)$.

5. Show that each pair of an elliptic involution are harmonic conjugates wo one other pair.[*]

[*] Holgate (Ref. 21, p. 213).

6. Prove that two involutions on a conic commute if and only if their centers are conjugate.*

7. Let $AA_1A'BB_1B'$ be a hexagon inscribed in a conic. Show how Pascal's theorem leads to a new proof of our theorem 4·68.

8. By 4·69 a given projectivity on a conic may be expressed as the product of two involutions. Let $O_i$ and $O_j$ be the two centers. Show that $O_i$ may be any point on the axis of the given projectivity and that $O_i$ is related to $O_j$ by a projectivity (on the axis) of the same type as the given projectivity (on the conic).

**7·6   A Generalization of Steiner's Construction.**   We have seen that the joins of corresponding points of an involution on a conic are concurrent. It is natural to ask what happens in the case of a projectivity that is not an involution.

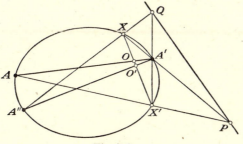

Fig. 7·6A

**7·61**   *The joins of corresponding points of two projectively related ranges on a conic envelop a conic (provided that the projectivity is not an involution).*

*Proof:*† Let $A$ be a fixed noninvariant point on the conic and $X$ a variable point on the conic, so that $AA'X \,\overline{\wedge}\, A'A''X'$, as in Fig. 7·6A. Then the axis of the projectivity is $PQ$, where

$$P = AX' \cdot XA', \qquad Q = A'X' \cdot XA''$$

Let $XX'$ meet the two fixed lines $AA'$ and $A'A''$ in $O$ and $O'$. From the quadrangle $AA'XX'$, $O$ and $P$ are conjugate points. From the quadrangle $A'A''XX'$, $O'$ and $Q$ are conjugate points. As $X$ and $X'$ vary on the conic, $P$ and $Q$ vary on the axis, $O$ varies on $AA'$, and $O'$ on $A'A''$;

* Veblen and Young (Ref. 42, p. 227).

† Due to James Jenkins while he was an undergraduate at the University of Toronto. For the complex version of this theorem see Baker (Ref. 2a, p. 52) or Coolidge (Ref. 5, p. 111). For another "real" proof see the *American Mathematical Monthly*, vol. 53, pp. 538–539, 1946.

thus

$$0 \underset{\wedge}{} P \overset{A}{\underset{\wedge}{}} X' \overset{A'}{\underset{\wedge}{}} Q \underset{\wedge}{} O'$$

But $O$ and $O'$ cannot coincide (as $A'$ is not an invariant point of the projectivity on the conic). Hence, by 6·58, the line $OO'$ (or $XX'$) envelops a conic.

### EXERCISES

1. Two sides of a variable triangle inscribed in a conic pass through fixed nonconjugate points. Prove that the third side envelops a conic (cf. Sec. 6·4, Exercise 3).

2. Those tangents to one conic which cut another conic determine on the latter an ordered correspondence. Show that this is not, in general, a projectivity. (*Hint:* Five arbitrary pairs of points on the conic may be related by such a correspondence; but of course no more than three arbitrary pairs can be related by a projectivity.)

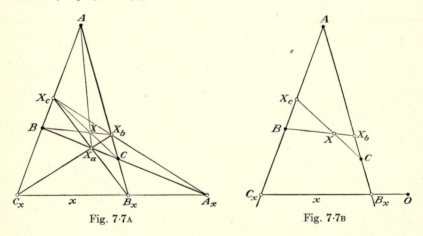

Fig. 7·7A            Fig. 7·7B

**7·7   Trilinear Polarity.**   We proceed to show how a triangle induces a correspondence between points not on its sides and lines not through its vertices. Although this is called a "trilinear polarity," it is not really a polarity at all; for, as we shall see, the "poles" of concurrent lines are not collinear points.

Given a triangle $ABC$ and a point $X$ not on any side, construct the six points

$$X_a = XA \cdot BC, \qquad X_b = XB \cdot CA, \qquad X_c = XC \cdot AB$$
$$A_x = X_b X_c \cdot BC, \qquad B_x = X_c X_a \cdot CA, \qquad C_x = X_a X_b \cdot AB$$

as in Fig. 7·7A. By Desargues' theorem for triangles $ABC$ and $X_aX_bX_c$, the three points $A_x$, $B_x$, $C_x$ lie on a line $x$. This is called the *trilinear polar* of $X$ (wo $ABC$).

Since $A_x$, $B_x$, $C_x$ are the harmonic conjugates of $X_a$, $X_b$, $X_c$ on the respective sides,* we can easily reconstruct $X$, the *trilinear pole* of a given line $x$ (not through any vertex).

Carrying over these definitions to the forbidden positions, we should say that the trilinear polar of a point on a side is that side itself, while the trilinear pole of a line through a vertex is that vertex itself. But the trilinear polar of a vertex and the trilinear pole of a side are indeterminate.

**7·71**　*The trilinear poles of a pencil of lines are the points of a conic circumscribing the triangle.*

*Proof:* Let $x$ be a variable line through a fixed point $O$, not on a side of the given triangle $ABC$, and $X$ its trilinear pole, as in Fig. 7·7B. From the involutions $(CC)(AA)$ and $(AA)(BB)$, we have

$$X_b \; \underset{\wedge}{} \; B_x \; \overset{O}{\underset{\overline{\wedge}}{=}} \; C_x \; \underset{\wedge}{} \; X_c$$

and hence the locus of $X = BX_b \cdot CX_c$ is a conic through $B$, $C$, and similarly through $A$. (This is called the *polar conic* of $O$ wo the triangle.)

If $O$, instead of being a point of general position, lies on a side, say on $BC$, then we have $X_b \; \overline{\underset{\wedge}{=}} \; X_c$, and the conic degenerates into that side and a line through the opposite vertex. Finally, if $O$ is taken at a vertex, the locus consists of the two sides through that vertex.

### EXERCISES

1. Show that the line $A_1B_1C_1$ of Fig. 2·4c is the trilinear polar of the point $S$.

2. Prove that the dual of the above construction for $x$ (given $X$) provides a construction for $X$ (given $x$). (*Hint:* Show that the three lines $AX$, $BB_x$, $CC_x$ are concurrent.)

3. Dualize 7·71.

4. If the trilinear polars of the vertices of a triangle (not $ABC$) are concurrent, prove that the trilinear poles of the sides are collinear.

5. Given a conic and an inscribed (or circumscribed) triangle, show that there is just one point whose polar wo the conic coincides with its polar wo the triangle. (*Hint:* Use Sec. 6·3, Exercise 3.)

6. Given an elliptic polarity and a self-polar triangle, show that there are just four points whose polars coincide with their trilinear polars. (*Hint:* Apply Sec. 7·5, Exercise 5, to the involutions of conjugate points on two sides of the triangle.)

---

* Poncelet (Ref. 30, p. 34) used this property to *define* the trilinear polar.

# CHAPTER 8

# AFFINE GEOMETRY

Projective geometry, in marked contrast to ordinary Euclidean geometry, is not at all concerned with length or distance; it contains no criterion for telling whether two segments are "congruent." But affine geometry takes us halfway back to the concept of distance: we are able to measure lengths along one line or on parallel lines and even to measure area, but we still cannot compare segments in different directions.

It is remarkable how many of the concepts and properties ordinarily considered in Euclidean geometry are still valid in the wider system of affine geometry. As we saw in Sec. 1·2, such concepts and properties are just those which are invariant under parallel projection.

Klein treated affine geometry by means of coordinates, viz., oblique Cartesian coordinates with independent scales of measurement along the two axes. The details of the synthetic treatment were worked out by Veblen.

**8·1  Parallelism.**  Affine geometry can be derived from projective by singling out one line $o$ and calling it the *line at infinity*, so as to be able to define parallelism. Any point on $o$ is called a *point at infinity*, and two lines are said to be *parallel* if their intersection is such a point. Strictly, affine geometry is concerned only with "ordinary" points and lines (not at infinity); hence we may say that the affine plane is derived from the projective plane by *removing* the line $o$. Thus two lines are parallel if they have no (ordinary) intersection. The theory could be built up in terms of ordinary points and lines alone; but we shall find it easier to make use of $o$, that is, to develop affine geometry from the projective point of view. However, all our theorems will be stated in terms of ordinary points and lines.

It is an immediate consequence of our definition of parallelism that just one line can be drawn, through a given point, parallel to a given line (not passing through the point), and that all lines parallel to a given line are parallel to one another.

Four points are said to form a *parallelogram OACB* if $OA$ is parallel to $BC$ and $OB$ to $AC$.

## *EXERCISES*

1. Let $OAMA'$ and $OBLB'$ be two parallelograms having their sides at $O$ along the same lines ($B$ on $OA$, $B'$ on $OA'$). Prove that the lines $AB'$, $BA'$, $LM$ are concurrent. (*Hint:* Use 4·31.)

2. Through a point $X$ draw a line $AC$ with $A$ and $C$ on two fixed parallel lines. Through a fixed point $O$ (not on any of these lines) draw $OA$, and let the parallel line through $C$ meet $OX$ in $X'$. Prove that the position of $X'$ is independent of the choice of the transversal $AC$ and that the correspondence $X \rightarrow X'$ is a homology.* (*Hint:* Figure 5·2B with $A'$ at infinity.)

Fig. 8·2A            Fig. 8·2B

**8·2  Intermediacy.** In Sec. 3·1 we considered a closed line on which the order of points is *cyclic*. The removal of a point at infinity changes this into an *open* line on which the order of points is *serial*, like the order of real numbers. In other words, separation is replaced by intermediacy: we say that $B$ is *between* two given points $C$ and $D$ when $AB//CD$, where $A$ is the point at infinity on the line $CD$. Axioms 3·11 to 3·16 lead to familiar properties of intermediacy:

**8·21** *For any two distinct points $B$ and $C$, there is at least one point $D$ such that $B$ is between $C$ and $D$.*

**8·22** *If $B$ is between $C$ and $D$, then $B$, $C$, $D$ are distinct.*

**8·23** *Any point between $C$ and $D$ is also between $D$ and $C$.*

**8·24** *Of any three collinear points, one is between the other two.*

**8·25** *If $B$ is between $C$ and $D$, while $C$ is between $B$ and $E$, then $B$ is between $D$ and $E$ (as in Fig. 8·2A).*

**8·26** *Intermediacy is preserved by parallel projection (Fig. 8·2B).*

If we wished to build up affine geometry as an independent system, instead of deriving it from projective geometry, we should take some of the above properties (along with certain statements about incidence) as a new set of axioms. Such a system is in some respects simpler than projective geometry; for *point* and *intermediacy* are the only primitive concepts needed: "line" and "incidence" can be defined in terms of them.

---

* La Hire, as quoted by Lehmer (Ref. 23, p. 110).

In affine geometry, any two points $B$ and $C$ determine a unique segment $BC$, namely, $BC/A$, where $A$ is the point at infinity on the line $BC$. This simply consists of all points between $B$ and $C$. Similarly, the *interior* of a triangle $ABC$ is the region $ABC/o$, in the notation of Sec. 3·8.

Although an ordinary line is open, the line at infinity is still closed. Thus we cannot say that one of three points at infinity lies between the other two, but only that two of four points at infinity separate the other two. It is important to notice that a positive sense of rotation is determined at all (ordinary) points simultaneously by calling one of the two senses along $o$ the positive sense.

### EXERCISES

1. Deduce from 3·31 that the affine line is decomposed by any one of its points into two half-lines, or *rays*. If $B$ lies between $C$ and $D$, the two rays are naturally denoted by $B/C$ and $B/D$. (Of course, $B/C$ is the one that does *not* contain $C$.)

2. Develop the affine theory of sense, using a symbol $S(BC) \neq S(CB)$.

**8·3  Congruence.*** Two segments $QQ'$ and $RR'$ are said to be *congruent by translation* if $QQ'R'R$ is a parallelogram. We then write

$$QQ' \equiv RR'$$

Theorem 1·51 with $O$ at infinity shows that the two relations $QQ' \equiv RR'$

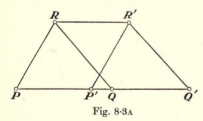

Fig. 8·3A

and $RR' \equiv PP'$ imply $PP' \equiv QQ'$. We naturally extend this notion so as to allow $PP'$ and $QQ'$ to be collinear, as in Fig. 8·3A: we write $PP' \equiv QQ'$ whenever there is a segment $RR'$ such that $PP'R'R$ and $QQ'R'R$ are parallelograms. It follows from this extended definition that the relation "congruent by translation" is reflexive ($PP' \equiv PP'$), symmetric (so that $PP' \equiv QQ'$ implies $QQ' \equiv PP'$), and transitive (so that $PP' \equiv QQ' \equiv RR'$ implies $PP' \equiv RR'$).

We proceed to show how the relation between congruent segments on one line may be expressed as a projectivity. In Fig. 4·4C we constructed the parabolic projectivity $MMA \barwedge MMA'$ by choosing points $R$ and $S$ on an arbitrary line through the invariant point $M$.

* Veblen and Young (Ref. 43, p. 75).

If $M$ is at infinity, we may take $RS$ to be the line at infinity, so that $AXX_0A_0$ and $A'X'X_0A_0$ are parallelograms and $AX \equiv A'X'$. Conversely, if $AX$ and $A'X'$ are congruent segments on the same line, we have parallelograms $AXX_0A_0$ and $A'X'X_0A_0$ from which we can reconstruct the figure for a parabolic projectivity. By 4·72, the relation $MMAX \mathrel{\overline{\overline{\wedge}}} MMA'X'$ implies $MMAA' \mathrel{\overline{\overline{\wedge}}} MMXX'$; that is,

$$AX \equiv A'X' \qquad \text{implies} \qquad AA' \equiv XX'$$

Hence:

**8·31**  *Two segments on the same line are congruent if and only if one is transformed into the other by a parabolic projectivity whose invariant point is at infinity; and a variable segment $XX'$ on a fixed line remains congruent to a fixed segment $AA'$ if and only if $X$ and $X'$ are related by such a projectivity.*

If $AA' = A'A''$, we call $A'$ the *mid-point* of the segment $AA''$. By 4·43, this means that $\mathrm{H}(MA', AA'')$, where $M$ is the point at infinity on the line $AA''$. Hence, writing $B$ for $A''$,

**8·32**  *The mid-point of a segment $AB$ is the harmonic conjugate (wo $A$ and $B$) of the point at infinity on the line $AB$.* *

### EXERCISES

1. Consider what happens to Fig. 2·5A when $CPQ$ is the line at infinity. Show that the diagonals of a parallelogram have the same mid-point. Deduce that the congruence $PP' = QQ'$ (on one line or on parallel lines) is equivalent to the statement that $PQ'$ and $QP'$ have the same mid-point.

2. Prove that the medians of a triangle are concurrent.* (*Hint:* See Sec. 7·7.)

**8·4  Distance.**  On a given line $AA_1$, let further points $A_2$, $A_3$, . . . be taken so that

$$AA_1 \equiv A_1A_2 \equiv A_2A_3 \equiv \cdots$$

Then we say that the *distance* $AA_n$ is $n$ times $AA_1$. Figure 8·4A illustrates both the parabolic projectivity

$$AA_1A_2 \ldots M \mathrel{\overset{R}{\overline{\wedge}}} BB_1B_2 \ldots M \mathrel{\overset{S}{\overline{\wedge}}} A_1A_2A_3 \ldots M$$

and the harmonic relations $\mathrm{H}(MA_1, AA_2)$, $\mathrm{H}(MA_2, A_1A_3)$, . . . . . This gives us a rule for multiplying a given distance by any positive integer. Can we also construct a fraction of a given distance? Not immediately, but soon.

* von Staudt (Ref. 40, p. 204, §338).

It is intuitively obvious that the segments $A_1A_2$, $A_2A_3$, . . . cover the whole ray $A_1/A$ (or $A_1M/A$), so that the point at infinity $M$ is the limit of the sequence of $A$'s and may appropriately be called $A_\infty$. But the rigorous proof of this fact is postponed till Sec. 10·2.

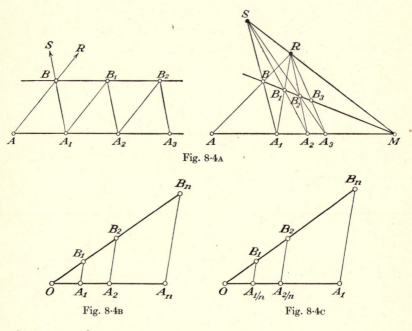

Fig. 8·4A

Fig. 8·4B                    Fig. 8·4c

Suppose we have

$$OA_1 \equiv A_1A_2 \equiv A_2A_3 \equiv \cdots$$

on one line and

$$OB_1 \equiv B_1B_2 \equiv B_2B_3 \equiv \cdots$$

on another, as in Fig. 8·4B. Then, because the successive $A$'s and $B$'s were constructed by taking harmonic conjugates, we have

$$OA_1A_nA_\infty \barwedge OB_1B_nB_\infty$$

By 4·23, this projectivity is a perspectivity whose center lies on the line at infinity $A_\infty B_\infty$. Hence $A_nB_n$ is parallel to $A_1B_1$, and the lines $A_nB_n$ for various $n$'s are parallel to one another.

Now, to divide a given segment $OA_1$ into $n$ equal parts, draw any other line through $O$, and take on it points $B_1$, $B_2$, . . . , $B_n$ so that

$$OB_1 \equiv B_1B_2 \equiv \cdots \equiv B_{n-1}B_n$$

Join $A_1B_n$, and draw parallel lines through the other $B$'s to meet $OA_1$ in points $A_{1/n}, A_{2/n}, \ldots, A_{(n-1)/n}$, as in Fig. 8·4c. Then

$$OA_{1/n} \equiv A_{1/n}A_{2/n} \equiv \cdots \equiv A_{(n-1)/n}A_1$$

and each of these segments is $1/n$ of $OA_1$.

In this manner we can construct a segment $OA_{m/n}$ for any given fraction $m/n$, and it is not difficult to see that the order of such points $A_{m/n}$ agrees with the order of the rational numbers $m/n$. (We naturally use $A_0$ as an alternative symbol for $O$.) Considerations of continuity* then enable us to define $A_x$ for any positive number $x$ (whether rational or not).

Negative numbers may be included by defining $A_{-x}$ as the harmonic conjugate of $A_x$ wo $A_0$ and $A_\infty$, so that $A_{-x}A_0 \equiv A_0A_x$; and then we write $A_0A_{-x} = -A_0A_x$, or $A_0A_x + A_0A_{-x} = 0$. By 4·63, $A_x$ and $A_{-x}$ are a typical pair of the involution $(A_0A_0)(A_\infty A_\infty)$.

Instead of saying that the distance $A_0A_x$ is $x$ times $A_0A_1$, we may say that the *ratio* $\dfrac{A_0A_x}{A_0A_1}$ is equal to the real number $x$. The above remarks provide a definition for the ratio of any two segments on one line or on parallel lines.† But it is important to realize that the ratio of two segments in any other relative position is essentially indeterminate. If $OAB$ is a triangle, the symbol $\dfrac{OB}{OA}$ has no numerical value (in affine geometry): we cannot say whether $OA$ or $OB$ is "longer." What we can say about segments on intersecting lines is as follows:

**8·41**   If $A'$ is on $OA$ and $B'$ on $OB$, with $A'B'$ parallel to $AB$, then $\dfrac{OA'}{OA} = \dfrac{OB'}{OB}$. *Conversely, if* $\dfrac{OA'}{OA} = \dfrac{OB'}{OB}$, $A'B'$ *must be parallel to* $AB$.

We are now ready to reconcile Desargues' treatment of involutions with von Staudt's. We have already considered the trivial case when the involution has an invariant point at infinity. In any other case the point at infinity is paired with an ordinary point $C$, called the *center* of the involution (though it has no connection with the $O$ of 7·52).‡

---

* The details are omitted because for this purpose Axiom 3·51 is quite unsuitable. The relevant treatment of continuity, based on the ideas of Weierstrass and Cantor, will be found in Chap. 10.

† Veblen and Young (Ref. 43, p. 85). The transitivity of the equality of ratios is a consequence of 2·25 in the special form 1·51.

‡ Young (Ref. 45, pp. 98–99).

When $RS$ is the line at infinity, Fig. 4·7A reduces to Fig. 8·4D, with $AQ$ parallel to $BP$ and $A'P$ to $B'Q$. By 8·41,

$$\frac{CA}{CB} = \frac{CQ}{CP} = \frac{CB'}{CA'}$$

that is, $CA \times CA' = CB \times CB'$. Hence:

**8·42**  *If $M$ is the point at infinity on the line $AA'$, the involution $(AA')(CM)$ relates points $X$ and $X'$ such that the product $CX \times CX'$ is constant.*

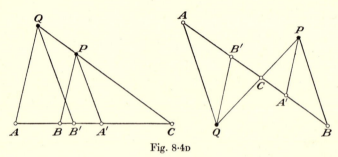

Fig. 8·4D

Strictly, a product of distances is not defined. (Certainly we must not follow Euclid in speaking of $CX \times CX'$ as a "rectangle"!) But the above statement is easily expressed in terms of ratios: we mean that $\dfrac{CX}{A_0A_1} \times \dfrac{CX'}{A_0A_1}$ is constant.

By 4·63, the center of a hyperbolic involution is midway between the two invariant points. Hence:

**8·43**  *The relation $\mathrm{H}(AB, XX')$ is equivalent to $CX \times CX' = CA^2$, where $C$ is the mid-point of $AB$.*

A "trivial" involution, for which the point at infinity is invariant, has no center in the above sense. But now another point takes over that role: the second invariant point $A_0$. A pair of corresponding points, being harmonic conjugates wo $A_0$ and $A_\infty$, are equidistant from $A_0$ on opposite sides, *i.e.*, the algebraic sum of their distances from $A_0$ is zero. More generally, taking an arbitrary origin $O$ instead of $A_0$, the sum of their distances from $O$ is constant.

### EXERCISES

1. On a line that meets the sides of a triangle $PQR$ in $A$, $B$, $C$, points $A'$, $B'$, $C'$ are chosen so that the segments $AA'$, $BB'$, $CC'$ all have the same mid-point. Prove that the lines $PA'$, $QB'$, $RC'$ are concurrent. (*Hint:* See 4·71.)

2. Prove that the mid-points of two pairs of opposite sides of a quadrangle form a parallelogram (whose sides are parallel to the remaining two sides of the quadrangle). Deduce that the three lines joining the mid-points of opposite

sides are concurrent (having, in fact, a common mid-point; see Sec. 8·3, Exercise 1).

3. Show that the relation $\mathrm{H}(AB, XX')$ is equivalent to

$$\frac{AX}{BX} + \frac{AX'}{BX'} = 0.$$

*Hint:* Evaluate $(CX - CA)(CX' - CB) + (CX - CB)(CX' - CA)$.

4. Show that the relation $\mathrm{H}(AB, XX')$ is equivalent to

$$\frac{1}{AX} + \frac{1}{AX'} = \frac{2}{AB},$$

so that $AB$ is the *harmonic mean* between $AX$ and $AX'$. (This is the origin of the name *harmonic set*.)

**8·5　Translation and Dilatation.**\* Let $o$ denote the line at infinity. An elation with axis $o$ is called a *translation*, and a homology with axis $o$ is called a *dilatation*. In particular, a harmonic homology with axis $o$ (that is, an involutory dilatation) is a *half-turn*. In this case, by 8·32, the joins of pairs of corresponding points all have the same mid-point; consequently the half-turn is sometimes called *reflection in a point* or *central inversion*. (We may think of it as the transformation that interchanges the numerals 6 and 9.)

In the first part of Fig. 5·2A (with $o$ at infinity), $ABB'A'$ is a parallelogram. Hence $AB = A'B'$ if and only if there is a translation that takes $AB$ to $A'B'$. By 5·21, the translation is determined when we are given the corresponding points $A$ and $A'$. We naturally call it the *translation from $A$ to $A'$*. The transitivity of the congruence relation is now revealed as a consequence of the fact that the product of the translation from $P$ to $Q$ and the translation from $Q$ to $R$ is the translation from $P$ to $R$ (see 5·25). The projectivity of 8·31 is induced by the translation in accordance with 5·23.

In this manner the notion of congruence can be extended from segments to figures of any kind: two figures are *congruent* if one can be derived from the other by a translation.

Similarly, two figures are said to be *homothetic* (or *similar and similarly situated*) if one can be derived from the other by a dilatation; e.g., two incongruent segments $AB$ and $A'B'$ on parallel lines are homothetic from the center $AA' \cdot BB'$. If instead the two incongruent

---

\* Veblen and Young (Ref. 43, pp. 92–95). A remarkably simple self-contained treatment of affine geometry (using 8·51 to *define* dilatation) has been given by Emil Artin, Coordinates in Affine Geometry, *Reports of a Mathematical Colloquium* (*Notre Dame*), 2·2, pp. 15–20, 1940.

segments are both on the same line whose point at infinity is $M$, they are related by the hyperbolic projectivity $MAB \overline{\wedge} MA'B'$. This is induced (according to 5·23) by a dilatation whose center is the second invariant point of the projectivity, *i.e.*, the center of the involution $(AB')(BA')$ (see the preamble to 4·67). Hence:

**8·51**   *Any two incongruent segments, on one line or on parallel lines, are related by a dilatation.*

In particular, if $AB \equiv B'A'$, the dilatation from $AB$ to $A'B'$ is a half-turn, and the center is the common mid-point of $AA'$ and $BB'$. This suggests the desirability of extending the meaning of congruent to include *congruent by a half-turn*, so that we can write

$$AB \equiv BA$$

From now on we shall use congruent in this wider sense, which will cause no confusion since, by 5·28, any translation can be expressed as the product of two half-turns. Accordingly, instead of the distance $AB$, which may be positive or negative according to the sense, we consider the *length* $AB$, which is essentially positive. Then two segments are congruent if and only if they have the same length.

By 5·24, if the sides of one triangle are parallel to respective sides of another, the two triangles are either homothetic or congruent. They are congruent by translation if the joins of corresponding vertices are parallel, and congruent by a half-turn if the joins have a common mid-point.

### EXERCISES

1. Show that the product of a translation and a half-turn is a half-turn.

2. Let $C$ and $C'$ be arbitrary points on the opposite sides $AB$ and $A'B'$ of a parallelogram $ABB'A'$. Let the line $(BC' \cdot CB')$ $(CA' \cdot AC')$ meet $AA'$ in $P$ and $BB'$ in $Q$. Prove that $AP \equiv QB'$. (*Hint:* Use 4·31 and the symmetry of the parallelogram.)

**8·6   Area.**\*   In affine geometry we cannot compare lengths in different directions. But we can compare areas in any position, since the ratio of two areas is invariant under parallel projection (see Sec. 1·2).

A region of the affine plane (not including any point at infinity) is called a *polygon* if it is entirely bounded by line segments. Clearly, any polygon can be dissected into a finite number of triangles. Two poly-

---

\* This is essentially the treatment of Hilbert (Ref. 19, Chap. IV), simplified by admitting continuity, but generalized by avoiding the use of right angles. For a more rigorous treatment see Veblen and Young (Ref. 43, pp. 96–104).

gons are said to be *equivalent* if (1) they can be dissected into a finite number of pieces that are congruent in pairs, or if (2) it is possible to annex to them one or more congruent pieces so that the completed polygons are equivalent in the first sense. In other words, two polygons are equivalent if they can be derived from each other by addition or subtraction of congruent pieces. By superposing two different dissections, we see that two polygons equivalent to the same polygon are equivalent to each other.

The parallelograms $OPRQ$ and $OPR'Q'$ of Fig. 8·6A are equivalent since the same trapezoid $OPR'Q$ is obtained by annexing triangle $PRR'$ to the former or $OQQ'$ to the latter. Hence:

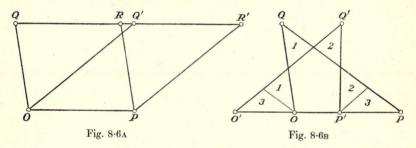

Fig. 8·6A                    Fig. 8·6B

**8·61**   *Two parallelograms are equivalent if they have one pair of opposite sides of the same length lying on the same pair of parallel lines.*

We could almost as easily prove this by taking the two parallelograms to have the same center (instead of a common base), so that the dissection would be centrally symmetrical. Since a parallelogram can be dissected along a diagonal into two triangles that are congruent by a half-turn, it follows that two triangles are equivalent if they have congruent sides on one line and opposite vertices on a parallel line (*i.e.*, equal bases and equal altitudes). The actual dissection for such a pair of triangles is illustrated in Fig. 8·6B.

By annexing a further triangle $OPQ'$ to each of two equivalent triangles with a common base $PQ'$, as in Fig. 8·6c, we deduce the following:

**8·62**   *Two triangles $OPQ$ and $OP'Q'$, having a common angle at $O$, are equivalent if the lines $PQ'$ and $P'Q$ are parallel.*

By "doubling" these triangles, as in Fig. 8·6D, we deduce:

**8·63**   *Two parallelograms $OPRQ$ and $OP'R'Q'$, having a common angle at $O$, are equivalent if the lines $PQ'$ and $P'Q$ are parallel.*

This suggests the propriety of selecting a certain parallelogram $OACB$ as unit of measurement and defining the *area* of a parallelogram

*OPRQ*, with *P* on *OA* and *Q* on *OB*, to be the number

$$\frac{OP}{OA} \times \frac{OQ}{OB}$$

By 8·41 and 8·63, two such parallelograms *OPRQ* and *OP'R'Q'* are equivalent if $\dfrac{OP'}{OP} = \dfrac{OQ}{OQ'}$, *i.e.*, if they have the same area. It follows that a parallelogram *OPRQ*, having two sides along the lines *OA* and

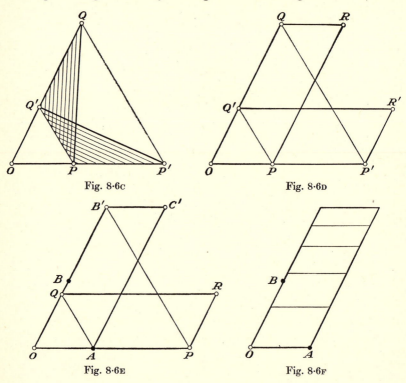

Fig. 8·6c                    Fig. 8·6d

Fig. 8·6e                    Fig. 8·6f

*OB*, is equivalent to a parallelogram *OAC'B'* where *B'* is constructed by drawing *PB'* parallel to *AQ*, as in Fig. 8·6e. The area is simply $\dfrac{OB'}{OB}$.

We now define the area of any polygon to be the area of an equivalent parallelogram *OAC'B'*. To see that the combined area of two or more juxtaposed polygons is the sum of their areas, we merely have to "stack" the equivalent parallelograms, as in Fig. 8·6f.

When we wish to compute the area of a given polygon, we first dissect it into convenient pieces and then draw an equivalent parallelo-

gram for each piece. A practical way to do this is to draw through each
vertex of the polygon a line parallel to $OA$, thus dissecting the polygon
into triangles, parallelograms, and trapezoids. We then dissect each
trapezoid into two triangles by drawing a diagonal. Each piece is now
a parallelogram or triangle having one side parallel to $OA$, and this
can be translated to a position where the side proceeds from $O$ along
$OA$. This in turn may be replaced
by a parallelogram with another side
along $OB$, using 8·61 if the piece is
a parallelogram and the following
device if it is a triangle: For a triangle
$OPS$ whose side $OP$ lies along $OA$, as
in Fig. 8·6G, an equivalent parallelogram with sides in the desired directions is $OPRQ$, where the line $QR$
joins the mid-points of $OS$ and $PS$, $Q$
lies on $OB$, and $PR$ is parallel to $OB$.

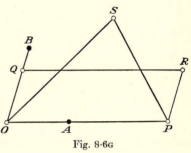

Fig. 8·6G

Finally, we alter the base $OP$ (unless $P$ already coincides with $A$) as in
Fig. 8·6D.

In this manner the ordinary properties of area may be established
without using any concepts (such as right angles) that are outside the
domain of affine geometry. The unit parallelogram $OACB$ takes the
place of the familiar unit square.

### EXERCISES

1. If the linear dimensions of a figure are doubled by a dilatation, show that
the area is quadrupled.

2. Give an affine proof for the following special case of Pappus's theorem:
If alternate vertices of a hexagon lie on two intersecting lines, while two pairs
of opposite sides are parallel, then the remaining sides are parallel.* (*Hint:* In
the notation of Fig. 4·3A with $LN$ at infinity, we have equivalent triangles
$OAA'$, $OBB'$, $OCC'$, by 8·62.)

**8·7 Classification of Conics.†** The line at infinity, $o$, enables us
to distinguish three types of conic: ellipse, parabola, and hyperbola.
By definition, the conic is an *ellipse* if $o$ is an exterior line (or *e* line), a
*parabola* if $o$ is a tangent, a *hyperbola* if $o$ is a secant (or *h* line). These
are shown diagrammatically in Fig. 8·7A. By the dual of 6·56, a unique
parabola can be drawn to touch the sides of a given quadrilateral.

The pole of $o$ is the *center*, $O$. Thus the center of an ellipse is an

---

* Veblen and Young (Ref. 43, p. 103); cf. Pasch and Dehn (Ref. 27, p. 226).

† von Staudt (Ref. 40, pp. 138, 205, §§248, 341–343).

interior point (*E* point), the center of a parabola is its point of contact with *o*, and the center of a hyperbola is an exterior point (*H* point). In the last case we can, of course, draw two tangents from the center: these are the *asymptotes* of the hyperbola (cf. Sec. 1·5, Exercise 2). Their points of contact are on the line at infinity and decompose the hyperbola into two arcs, called the two *branches*.

Any line through *O* is a *diameter*. Since the center of an ellipse is interior, all its diameters are secants. Since the center of a parabola is at infinity, its diameters are parallel secants. Since the center of a hyperbola is exterior, it has diameters of all kinds: the two asymptotes

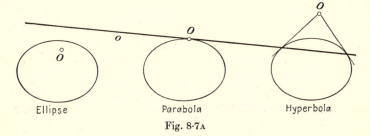

Ellipse                    Parabola                    Hyperbola

Fig. 8·7A

separate those which are secants from those which do not meet the hyperbola at all.

In the domain of ordinary points, a parabola has no center. Accordingly the ellipse and hyperbola are called *central* conics.

When we apply 6·83 to the line at infinity, we find that the nine-point conic contains the mid-points of the sides of the quadrangle and the centers of all the conics of the quadrangular pencil. Hence:

**8·71**   *The mid-points of the six sides of a quadrangle and the three diagonal points all lie on a conic.*

### EXERCISES

1. Show that a central conic is symmetrical about its center. (Apply Exercise 4 of Sec. 6·4 to the half-turn about *O*.)

2. If a diameter of a central conic is a secant, show that the tangents at its two "ends" are parallel. On the other hand, a parabola has no two parallel tangents.

3. Prove that the locus of centers of conics inscribed in a quadrilateral is a line through the mid-points of the three diagonals. (*Hint:* Use the dual of 5·81.) Deduce that this line is a diameter of the escribed parabola.

4. Show that the three lines joining the mid-points of opposite sides of a quadrangle are diameters of the nine-point conic (cf. Sec. 6·8, Exercise 5, and Sec. 8·4, Exercise 1).

5. Justify the construction of Sec. 6·5, Exercise 6.

6. Prove that the center of a conic cannot be interior to a self-polar triangle. (*Hint:* Use 6·21.)

7. Through a variable point $X$ on a fixed line, a line $c$ is drawn parallel to the polar of $X$ wo a given ellipse. Prove that $c$ envelops a parabola.*

**8·8  Conjugate Diameters.**† A *chord* of a conic is the segment joining two distinct points on the conic.

Fig. 8·8A

**8·81**  *The mid-points of parallel chords lie on a diameter.*

*Proof:* The parallel chords have a common point at infinity whose polar bisects them all, by 6·41 and 8·32. This polar, being conjugate to $o$, is a diameter conjugate to all the chords.

Hence, to construct a diameter, we merely have to join the mid-points of two parallel chords, as in Fig. 8·8A. The center can be found as the point of intersection of two diameters.‡

By 5·53, conjugate diameters of a central conic are pairs of an involution. Since the invariant lines (if any) are asymptotes, this involution is elliptic or hyperbolic according as the conic is an ellipse or a hyperbola. (In fact, this is the origin of the names for the two types of involution, and thence, by analogy, the names *elliptic, parabolic, hyperbolic* for the three types of projectivity.) By 6·42, as La Hire observed:

**8·82**  *Any pair of conjugate diameters of a hyperbola are harmonic conjugates wo the asymptotes.*

Two conjugate diameters form, with the line at infinity, a self-polar triangle, which has one $e$ side and two $h$ sides, as in Fig. 6·2c (second

* Chasles (Ref. 4, p. 138).

† Apollonius (Ref. 1, pp. 48–93, lib. I); Chasles (Ref. 4, pp. 114–117); Reye (Ref. 31, pp. 100–107).

‡ Apollonius (Ref. 1, pp. 265–267, lib. II, prop. 44, 45).

part). Hence, in the case of an ellipse both diameters are secants, but in the case of a hyperbola one is a secant and the other an exterior line.

We shall make use of the next theorem in Sec. 9·2, where we develop the theory of circles.

**8·83**  *If a parallelogram is inscribed in a conic, its two diagonals are diameters of the conic and its sides are parallel to a pair of conjugate diameters.*

*Proof:* The sides and diagonals of the parallelogram form a quadrangle whose diagonal triangle has two vertices at infinity, so one side of this triangle is *o*. By 6·43, its other two sides are conjugate diameters parallel to the sides of the parallelogram. (The conic, having the same center as the parallelogram, obviously cannot be a parabola.)

The following theorem makes a neat companion for 8·81:

**8·84**  *The mid-points of chords that pass through a fixed point (not on an asymptote) lie on another conic.*

*Proof:* Let $P$ be the point at infinity on the chord $x$ through the fixed point $A$. This chord is bisected by the conjugate diameter $p$, and we have $x \overline{\wedge} P \overline{\wedge} p$. The lines $x$ (through $A$) and $p$ (through $O$) can never coincide unless $OA$ is an asymptote. Hence, by 6·54, the point $x \cdot p$ lies on a certain conic through $A$ and $O$.

Theorem 5·82 shows that we can find infinitely many conics having a given involution of conjugate points on a line $a$ whose pole is a given point $A$. In particular, we can find infinitely many conics having a given center and a given involution of conjugate points on *o*, *i.e.*, having a given involution of conjugate diameters. If the involution is elliptic, the conics are ellipses any two of which are related by a dilatation (see 6·84). If instead it is hyperbolic, the conics are hyperbolas having the same asymptotes (*i.e.*, having double contact at infinity). They fall into two subsystems such that any two hyperbolas belonging to the same subsystem are related by a dilatation. But two belonging to different subsystems are not so simply related;* for they are separated by their common asymptotes. To derive one from the other we need a homology whose center is at infinity (*i.e.*, a transformation of the kind that von Staudt called an *affinity*). In the special case when this is a *harmonic* homology (or *affine reflection*) they are called *conjugate* hyperbolas. These remarks may be summarized as follows:

**8·85**  *A conic can be drawn through a given point so as to have two given pairs of concurrent lines as conjugate diameters. The conic will be an*

---

* It is amusing to observe that Chasles (Ref. 4, pp. 246, 248) missed this little complication, although many properties of conjugate hyperbolas had been known since the time of Apollonius.

*ellipse or a hyperbola according as the pairs do or do not separate each other. By varying the point we obtain, in the former case, a system of homothetic ellipses all having the same involution of conjugate diameters and, in the latter, two "conjugate" systems of homothetic hyperbolas all having the same asymptotes.*

### EXERCISES

1. Applying 8·81 to a hyperbola, investigate the nature of the diameter when the chords (*a*) join points on the same branch or (*b*) join a point on one branch to a point on the other. Observe that in the former case the diameter passes through the points of contact of two tangents parallel to the chords.

2. Prove that the diagonals of a parallelogram circumscribed about a central conic are conjugate diameters.

3. As a corollary of 8·83, any parallelogram inscribed in a conic is concentric with the conic. Deduce that the mid-points of the six sides of a quadrangle lie on a conic. (This provides an elementary proof for part of 8·71.)

4. Given two conjugate diameters *a*, *b* and a point *P* on the conic (but not on *a* or *b*), construct an inscribed parallelogram *PSQR* whose sides are parallel to *a* and *b*.

5. Show that, when 8·84 is applied to a parabola, the locus is another parabola.*

6. Let *aa'* and *bb'* be two pairs of conjugate diameters of an ellipse. Prove that the sides of the parallelogram formed by the ends of *a* and *b* are parallel to those of the parallelogram formed by the ends of *a'* and *b'*. (*Hint:* Let *m* and *n* be the common harmonic conjugates of the pairs *ab* and *a'b'*. The involution of conjugate diameters interchanges these pairs. Being elliptic, it cannot leave *m* and *n* separately invariant and hence must interchange them.)

**8·9  Asymptotes.**† The hyperbola provides a greater variety of special theorems than the other kinds of conic do, because of the existence of asymptotes. It is hoped that, after following the proofs of a few specimen theorems, the reader will feel prepared to deal with any affine properties of the hyperbola that may be proposed, either in the accompanying exercises or elsewhere.

**8·91** *Any tangent to a hyperbola meets the two asymptotes in points equidistant from the point of contact.*

*Proof:* If *M* is the point of contact of the tangent *PQ*, as in Fig. 8·9A, the parallel diameter is conjugate to *OM*. By 8·82, these two diameters are harmonic conjugates wo the asymptotes *OP* and *OQ*. Thus *M* and the point at infinity on *PQ* are harmonic conjugates wo *P* and *Q*; *i.e.*, *M* is the mid-point of *PQ*.

---

* Smith (Ref. 39, p. 137, No. 15).
† Apollonius (Ref. 1, pp. 198–231, lib. II, prop. 3–21).

**8·92**  *If a chord AB of a hyperbola meets the asymptotes in P and Q, then PA ≡ BQ.*

*Proof:* The chord $AB$ is bisected by the conjugate diameter, as we saw in the proof of 8·81 (see Fig. 8·9B). But 8·82 shows that this same diameter bisects the segment $PQ$. Thus $AB$ and $PQ$ have the same mid-point.

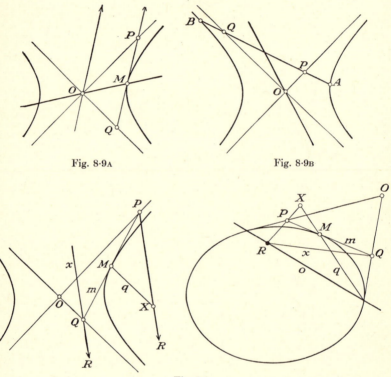

Fig. 8·9A          Fig. 8·9B

Fig. 8·9c

**8·93**  *If the tangent to a hyperbola at M meets the two asymptotes in P and Q, while MX is parallel to the latter asymptote, then the polar of X is the line through Q parallel to PX.*

*Proof:* Let $R$ denote the point at infinity on $PX$, as in Fig. 8·9c. This point is conjugate to the vertex $O$ of the triangle $OPQ$, which is circumscribed about the hyperbola. Hence, by 6·57, $PR$ and $QR$ are conjugate lines. Moreover, $MX$, joining the points of contact of the two tangents from $Q$, is the polar of $Q$. Thus $QR$, being conjugate to both $PR$ and $MX$, is the polar of their point of intersection, $X$.

**8·94**  *A variable tangent to a hyperbola cuts off from the asymptotes a triangle of constant area*

*Proof:* Let $PQ$ and $P'Q'$ be two tangents to the hyperbola, with $P$ and $P'$ on one asymptote, $Q$ and $Q'$ on the other, as in Fig. 8·9D. These

Fig. 8.9D

two tangents form, with the two asymptotes, a quadrilateral $OPP'QQ'R$, circumscribed about the hyperbola. The three diagonal lines are $PQ'$, $P'Q$, and $OR$. The last is a diameter; therefore its pole, $PQ' \cdot P'Q$, is at infinity, which means that $PQ'$ and $P'Q$ are parallel. By 8·62, triangles $OPQ$ and $OP'Q'$ have the same area.

**8·95**  Corollary: *If a variable line cuts off from two fixed lines a triangle of constant area, its envelope is a hyperbola, and the locus of the mid-point (of the segment intercepted) is the same hyperbola.**

### EXERCISES

1. If through any point $A$ a line $APB$ is drawn parallel to an asymptote of a hyperbola, cutting the curve in $P$ and the polar of $A$ in $B$, show that $P$ is the mid-point of $AB$.*

2. Prove that the mid-points of chords of a hyperbola that pass through a fixed point on an asymptote lie on a line parallel to the other asymptote. (*Hint:* Use 8·92.)

3. A variable segment has its ends on two fixed lines and passes through a

* Smith (Ref. 39, p. 203, Nos. 4 and 11).

fixed point. Prove that the locus of its mid-point is a hyperbola whose asymptotes are parallel to the given lines.*

4. If a tangent to a hyperbola meets the two asymptotes in $P$ and $Q$, prove that any two parallel lines through $P$ and $Q$ are conjugate.

5. A parallelogram has its sides parallel to the asymptotes of a hyperbola, and one of its diagonals is a chord. Prove that the other diagonal passes through the center of the hyperbola.*

6. Let $PQ$ and $P'Q'$ be two parallel tangents to a hyperbola, with $P$ and $P'$ on one asymptote, $Q$ and $Q'$ on the other. Prove that $PQ'$ and $P'Q$ are tangents to the conjugate hyperbola, and that the parallelogram $PQ'P'Q$ has constant area.

7. When a quadrangle is convex, so that its nine-point conic wo $o$ is a hyperbola, Exercise 5 of Sec. 6·8 shows that three pairs of the nine points are joined by diameters; therefore three points occur on one branch of the hyperbola while three take diametrically opposite positions on the other. Prove that the remaining three points (the diagonal points of the quadrangle) are all on one branch, *i.e.*, that the nine points are distributed as $3 + 6$ between the two branches. (This is not easy.)

* Smith (Ref. 39, p. 203, Nos. 3 and 12).

## CHAPTER 9

## *EUCLIDEAN GEOMETRY*

The time has come for us to fulfill the promise of Sec. 1·6, that we should return to ordinary geometry from a new point of view. We shall see how von Staudt's idea of choosing an elliptic involution on the line at infinity of the affine plane enables us to define perpendicularity and congruence, so that distances can be compared in any direction. Many problems of Euclidean geometry are most easily solved by the projective approach. But at this stage we are free to use either the new method or the old, whichever is found more convenient at the moment.

**9·1  Perpendicularity.**  We have seen that affine geometry can be derived from real projective geometry by singling out for special treatment a line "at infinity," which enables us to say when two other lines are parallel. Similarly, we shall find that Euclidean geometry can be derived from affine geometry by singling out an elliptic involution on that special line, to serve as the *absolute* involution, which enables us to say when two lines are perpendicular. From the standpoint of projective geometry, all elliptic involutions are exactly alike; but as soon as we have specialized one such involution, we can say (as a definition) that two lines shall be called *perpendicular* if their points at infinity are a pair of the absolute involution.

To see that this agrees with our usual ideas about perpendicularity, we merely have to observe that the correspondence between perpendicular lines is symmetric and preserves harmonic sets, so that it is an involution: elliptic because no line is perpendicular to itself. We call it the *orthogonal* involution.

Many properties of perpendicularity are immediate consequences of this definition; *e.g.*, two perpendiculars to one line are parallel. The perpendicular line through the mid-point of a given segment is called the *right bisector* of the segment. A parallelogram that has two perpendicular sides is called a *rectangle*. A triangle that has two perpendicular sides is called a *right triangle;* any other kind of triangle is said to be

117

*oblique.* The *altitudes* of a triangle are defined to be the perpendiculars from the vertices to the respectively opposite sides. The "feet" of the altitudes are said to form the *pedal* triangle.

**9·11**   *The three altitudes of a triangle are concurrent.*

*Proof:*\* For a right triangle this is trivial, so let us assume the given triangle *PQR* to be oblique, as in Fig. 9·1A. Let the altitudes from *P*

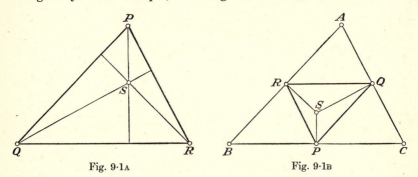

Fig. 9·1A                    Fig. 9·1B

and *Q* intersect in *S*. Then the quadrangle *PQRS* determines on the line at infinity a quadrangular set of points, two of whose pairs belong to the absolute involution. Hence, by 4·71, the third pair likewise belongs to this involution, and *RS* must be the third altitude of the triangle.

The point *S* is called the *orthocenter* of triangle *PQR*. (In the case of a right triangle, with *PR* perpendicular to *QR*, the orthocenter coincides with *R*.)

**9·12**   Corollary: *If S is the orthocenter of an oblique triangle PQR, then P is the orthocenter of QRS, and so on. The pedal triangle of any one of these four triangles is the diagonal triangle of the quadrangle PQRS.*

We define a *rectangular* hyperbola as one whose asymptotes are perpendicular (or whose points at infinity are a pair of the absolute involution). By Desargues' involution theorem (our 6·73):

**9·13**   *Every conic through the vertices and orthocenter of an oblique triangle is a rectangular hyperbola.*†

If *P*, *Q*, *R* are the mid-points of the sides of a triangle *ABC*, as in Fig. 9·1B, the right bisectors of those sides are the altitudes of the medial triangle *PQR*. Hence:

\* von Staudt (Ref. 40, p. 205, §339).

† Baker, (Ref. 2a, p. 83).

**9·14**   *The right bisectors of the three sides of a triangle are concurrent.*

The point of concurrence (which is the orthocenter of $PQR$) is called the *circumcenter* of $ABC$.

### EXERCISES

1. Prove that those chords of a conic which subtend a right angle at any fixed point on the conic are concurrent.*
2. Show that every rectangular hyperbola through the vertices of a triangle passes also through the orthocenter.
3. Prove that just one rectangular hyperbola can be drawn through the vertices of a quadrangle not consisting of a triangle and its orthocenter. (By Exercise 2, this passes also through the orthocenters of the four triangles determined by these vertices.†)

**9·2   Circles.**   In affine geometry there is no distinction of shape between one ellipse and another: though we can say which has the greater area, we cannot say which has the greater "eccentricity." But the absolute involution enables us to make this further distinction, and in particular to define a circle.

*Definition:* A *circle* is a conic for which the involution of conjugate diameters coincides with the involution of perpendicular diameters.

By applying 8·85 to two pairs of perpendicular lines through one point, we verify that such a conic exists. In fact:

**9·21**   *Just one circle may be described with any center to pass through any given point.*

Thus we have proved Euclid's third postulate. It follows that a unique circle can be drawn with any given segment for a diameter. When a circle is drawn with center $O$ to pass through a point $A$, the segment $OA$ is called a *radius*.

The diameter conjugate to a given chord is its right bisector. Conversely, any point $O$ on the right bisector of a given segment $AB$ is the center of a circle through $A$ and $B$; for the circle with radius $OA$ must pass through $B$ also, since the line $AB$ is conjugate to the right bisector. Hence the circumcenter of a triangle $ABC$ (9·14) is the center of a circle through the three vertices:

**9·22**   *Any three noncollinear points lie on a unique circle.*

This is called the *circumcircle* of the triangle.

---

* von Staudt (Ref. 40, p. 206, §344). The point of concurrence is known as the *Frégier point*. It lies on the normal to the conic at the given point.

† See Holgate (Ref. 21, p. 207).

Since any two conjugate diameters are perpendicular,

**9·23**    *The polar of a point wo a circle is perpendicular to the diameter through the point.*

In particular:

**9·24**    *A tangent to a circle is perpendicular to the diameter through its point of contact.*

Since the involution of perpendicular diameters is elliptic, *a circle is an ellipse.* Ellipses that are not circles may conveniently be called *eccentric* ellipses. Since an involution is determined by two of its pairs,

**9·25**    *If a conic has two distinct pairs of perpendicular conjugate diameters, it must be a circle.*

By 8·83, any parallelogram inscribed in a circle is a rectangle whose center is the center of the circle. Hence:

**9·26**    *The lines joining the ends of a diameter to any other point on the circle are perpendicular.*

In other words, "the angle in a semicircle is a right angle" (Euclid III. 31). Conversely,

**9·27**    *The locus of the point of intersection of perpendicular lines through two fixed points is a circle.*

*Proof:* Let $M$ and $N$ be the points at infinity on the two perpendicular lines $x$ and $y$. Since $x \mathbin{\overline{\wedge}} M \mathbin{\overline{\wedge}} N \mathbin{\overline{\wedge}} y$, the locus is a conic (6·53) that has infinitely many inscribed rectangles with two fixed opposite vertices. By 8·83 and 9·25, any two of these rectangles suffice to make the conic a circle.

By 8·85, concentric circles are homothetic. But we can say far more:

**9·28**    *Any two circles are either congruent or homothetic.*

*Proof:* We remark first that a translation or dilatation, being a collineation that leaves invariant every point at infinity, transforms perpendicular lines into perpendicular lines and circles into circles. Let the line joining the centers of two given circles determine respective diameters $AB$ and $CD$, and let $M$ be the point at infinity on this line. Applying 8·51, we see that if $AB \not\equiv CD$ (as in Fig. 9·2A), one of the circles is transformed into the other by a dilatation from the center of either of the involutions

$$(AD)(BC), \qquad (AC)(BD)$$

(These points, $N_1$ and $N_2$, are called the *centers of similitude* of the two circles.) But if $AB \equiv CD$, one of the dilatations has to be replaced by a translation, while the other becomes a half-turn.

Some of the above ideas suggest an elementary proof for the famous theorem of the nine-point circle:

**9·29**   *If S is the orthocenter of a triangle PQR, then the mid-points of the six segments QR, RP, PQ, PS, QS, RS and the feet of the three altitudes all lie on a circle.*

*Proof:*[*] Let $L, M, N, L', M', N'$ be the six mid-points and $PA, QB, RC$ the three altitudes, as in Fig. 9·2B. By 8·41, both $NM$ and $M'N'$

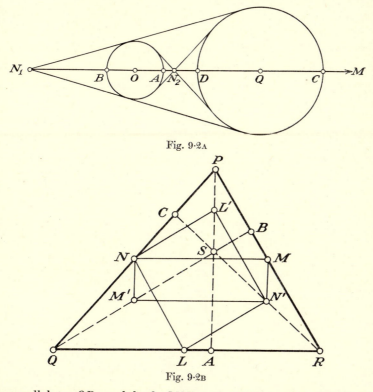

Fig. 9·2A

Fig. 9·2B

are parallel to $QR$, and both $MN'$ and $NM'$ parallel to $PS$. Thus $MNM'N'$ is a rectangle. Similarly $NLN'L'$ is a rectangle. Hence $LL'$, $MM'$, $NN'$ are three diameters of a circle. Since $LA$ and $L'A$ are perpendicular, this circle passes through $A$; similarly, it passes through $B$ and $C$.

It is interesting to observe that this is a special case of the nine-point conic (8·71). Thus the centers of the rectangular hyperbolas of 9·13 all lie on the nine-point circle.

[*] This proof, due to V. E. Dietrich, resembles that of Durell (Ref. 10, p. 27).

A certain dilatation from the orthocenter $S$ will transform the mid-points of $PS$, $QS$, $RS$ into the vertices $P$, $Q$, $R$ themselves and will consequently transform the nine-point circle of triangle $PQR$ (which is the circumcircle of $ABC$) into the circumcircle of $PQR$. In other words, the nine-point circle is the locus of a point midway between the orthocenter and a point that runs round the circumcircle. It follows that the nine-point center is midway between the orthocenter and circumcenter. The line on which these three points lie is called the *Euler line* of triangle $PQR$.

Theorems 9·26 and 9·27 show that the above definition for a circle is equivalent to Euclid's. Hence we can now reconcile von Staudt's definition for a conic with the classical one. For, any plane section of a circular cone is a central projection of a circle (as described in Sec. 1·3).

### EXERCISES

1. If a triangle is self-polar wo a circle, show that its orthocenter is the center of the circle. (*Hint:* Apply 9·23 to each vertex of the triangle.)

2. Prove that the *centroid* (Sec. 8·3, Exercise 2) lies on the Euler line.*

3. Prove that the centers of similitude are harmonic conjuates wo the centers of the two circles. (*Hint:* In the notation of Fig. 9·2A we have

$$N_1A \times N_1D = N_1B \times N_1C \qquad \text{and} \qquad N_2A \times N_2C = N_2B \times N_2D$$

therefore $N_1O/N_1Q = OA/QC = -N_2O/N_2Q$.)

4. Let $CA$ and $CB$ be two tangents to a circle with center $O$. Join $A$, $B$, $C$ to any point $S$ on the circle by lines cutting the diameter perpendicular to $OS$ in $A'$, $B'$, $C'$. Prove that $C'$ is the mid-point of $A'B'$. (*Hint:* $SA$ and $SB$ are harmonic conjugates wo $SC$ and the tangent at $S$.) This could have been made into an affine theorem by changing the words *circle* and *perpendicular* into *central conic* and *conjugate*. But the Euclidean theorem has an interesting application to solid geometry; for it shows that, when a small circle $AB$ on a sphere is projected stereographically from $S$, the center $C'$ of the projected circle comes from the vertex $C$ of the corresponding enveloping cone.†

**9·3  Axes of a Conic.** An *axis* of a conic is defined as a diameter which is perpendicular to the chords it bisects (*i.e.*, it is an axis of symmetry; see Fig. 9·3A). In the case of a parabola, all diameters are parallel, and there is just one system of chords perpendicular to them. Hence:

**9·31**  *A parabola has just one axis.*

This can be constructed by joining the mid-points of two chords

* Robinson (Ref. 32, p. 48).
† Donnay (Ref. 9, p. 10).

drawn perpendicular to any diameter* (and the diameter can be constructed by joining the mid-points of any two parallel chords).

In the case of a central conic (ellipse or hyperbola) we consider two involutions of diameters:

1. Conjugate diameters (an elliptic or hyperbolic involution).
2. Perpendicular diameters (always an elliptic involution).

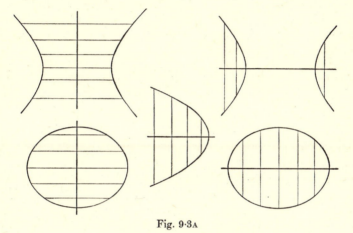

Fig. 9·3A

For a circle, these two involutions coincide. In every other case we can find (by the dual of 7·55) just one common corresponding pair. This is a pair of perpendicular conjugate diameters, *i.e.*, a pair of axes. Hence:

**9·32**  *A central conic, other than a circle, has just two axes.*

But a circle has infinitely many axes: every diameter is an axis.

The axes of an ellipse (like all its diameters) are secants. Likewise the axis of a parabola is a secant (though one of the intersections is at infinity). But in the case of a hyperbola, one axis is a secant while the other is an exterior line; these are called the *transverse* axis and *conjugate* axis, respectively.

When an axis is a secant, the points where it cuts the conic are called *vertices*. Thus a parabola has one vertex (unless we agree to admit another at infinity), a hyperbola has two vertices, and an eccentric ellipse has four.† But a circle has infinitely many: every point on the circle is a vertex.

* Apollonius (Ref. 1, pp. 267–271, lib. II, prop. 46–48).
† von Staudt (Ref. 40, p. 206, §342).

## EXERCISES

1. Show that the circle on the transverse axis of a hyperbola as diameter has double contact with the hyperbola and that the circle on either axis of an eccentric ellipse as diameter has double contact with the ellipse.

2. Consider a conic and a point $P$ not on it nor at its center (if it has a center). Let $y$ be the line through $P$ perpendicular to the chords bisected by a variable diameter $x$. Prove that the locus of $x \cdot y$ is a rectangular hyperbola through $P$. (This is known as Apollonius's hyperbola. It passes through the feet of the normals that can be drawn from $P$ to the given conic.)

**9·4  Congruent Segments.**  In affine geometry we were able to define congruent segments on one line or on parallel lines. In Euclidean geometry we can define congruent segments in different directions, as follows:

Segments $OA$ and $OB$ are said to be *congruent by rotation* if $A$ and $B$ lie on a circle with center $O$. Then we write

$$OA \equiv OB$$

(Radii of a circle are congruent.) If a translation takes $OA$ and $OB$ to two radii $O'A'$ and $O'B'$ of another circle (so that $OA \equiv O'A'$ and $OB \equiv O'B'$ by translation), we simply say that $OA$ and $O'B'$ are *congruent*, writing $OA \equiv O'B'$. This relation is clearly reflexive, symmetric, and transitive. The method of Sec. 8·4 now enables us to define the *length* of any segment in terms of a standard segment as unit. In particular, if $C$ is exterior to the circle with radius $OA$, we say that $OC$ is *greater* (or longer) than $OA$, writing

$$OC > OA$$

If two axes of an ellipse are congruent, their four ends (forming a square) lie on a circle that touches the ellipse at all four places and therefore coincides with it entirely. Hence *an ellipse having equal axes is a circle*. Apart from this case, one axis of the ellipse must be longer than the other. The longer and shorter are called the *major* and *minor* axes.

Conjugate points on a diameter of a circle are said to be *inverse* wo the circle. By 8·42, two inverse points have a constant product of distances from the center. Since any point on the circle is its own inverse, this product is equal to the square of the radius. In virtue of 9·23, we may express this result as follows:

**9·41**  *For a circle of radius $\rho$, the product of the central distances of a point and its polar is equal to $\rho^2$.*

Show that any affine property of a parallelogram can be deduced from the corresponding property of a square. Hence (or otherwise) prove the following affine theorem:

Of all the ellipses circumscribing a given parallelogram, the one for which the diagonals are *conjugate* diameters has the smallest area.

*Hint:* We know from elementary analytic geometry that the ellipse

$$\frac{x^2}{a^2} + \frac{y^2}{b^2} = 1$$

of area $\pi ab$, is circumscribed about the square $(\pm c, \pm c)$ if

$$\frac{1}{c^2} = \frac{1}{a^2} + \frac{1}{b^2} = \frac{2}{ab} + \left(\frac{1}{a} - \frac{1}{b}\right)^2$$

**9·5  Congruent Angles.** Let $a$ and $b$ be two intersecting lines, while $a'$ and $b'$ are the respective perpendiculars through their common point. Let $c$, $d$, $c'$, $d'$ be another such set of concurrent lines. The

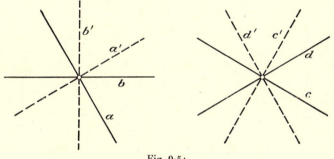

Fig. 9·5A

ordered pair of lines $\langle ab \rangle$ is called an *angle*. We say that the two angles $\langle ab \rangle$ and $\langle cd \rangle$ are *congruent* if the following two conditions are satisfied:

(i) $\qquad\qquad\qquad aba'b \mathbin{\overline{\wedge}} cdc'd'$

(ii) $\qquad\qquad\qquad S(aba') = S(cdc')$

(see Fig. 9·5A). We then write

$$\langle ab \rangle \equiv \langle cd \rangle$$

This is clearly an equivalence relation, like the congruence of segments. It holds if $a$ and $b$ are respectively parallel to $c$ and $d$ (in which case $aba'b' \overset{o}{\mathbin{\overline{\wedge}}} cdc'd'$). Moreover, $\langle ab \rangle \equiv \langle a'b' \rangle$.

Angles $\langle ab \rangle$ and $\langle ba \rangle$ are said to be *supplementary*. The angle $\langle aa' \rangle$ is called a *right* angle.

**9·51**   *The only angle congruent to its supplement is a right angle.*

*Proof:* Given $\langle ab \rangle \equiv \langle ba \rangle$, we wish to show that $b = a'$ (and consequently $b' = a$). If this were not so, we should have four distinct lines $a$, $b$, $a'$, $b'$ such that

$$S(aba') = S(bab') \neq S(abb')$$

whence $ab//a'b'$. But since the involution of perpendicular lines $(aa')(bb')$ is elliptic, we have $aa'//bb'$. This provides the desired contradiction.

If three concurrent lines $a$, $b$, $m$ have the property $\langle am \rangle \equiv \langle mb \rangle$, we say that $m$ *bisects* the angle $\langle ab \rangle$.

**9·52**   *If a line $m$ bisects $\langle ab \rangle$, then the perpendicular line $n$ does likewise, and* $\mathrm{H}(mn, ab)$.

*Proof:* We have $ama'n \underset{\wedge}{\phantom{x}} mbnb' \underset{\wedge}{\phantom{x}} b'nbm$, by the dual of 2·71. Therefore $mn$ is a pair of the involution $(ab')(a'b)$. It is also a pair of the orthogonal involution $(aa')(bb')$. Hence $m$ and $n$ are the invariant lines of the product of these two involutions, which is $(ab)(a'b')$. This proves that $\mathrm{H}(mn, ab)$. Moreover, $ana'm \underset{\wedge}{\phantom{x}} mb'nb \underset{\wedge}{\phantom{x}} nbmb'$, and, reversing the sense $S(ama') = S(mbn)$, we have $S(ana') = S(nbm)$; therefore $\langle an \rangle \equiv \langle nb \rangle$.

It follows that the relation $\langle am \rangle \equiv \langle cm \rangle$ for concurrent lines $a$, $m$, $c$ implies $a = c$; therefore the same relation for nonconcurrent lines implies that $a$ and $c$ are parallel.

The converse of 9·52 was discovered by Desargues:[*]

**9·53**   *If $\mathrm{H}(mn, ab)$ and $m$ is perpendicular to $n$, then $m$ and $n$ bisect the angles between $a$ and $b$.*

*Proof:* Applying the orthogonal involution to the given harmonic set, we have $\mathrm{H}(nm, a'b')$. Thus $ab$ and $a'b'$ are two pairs of the involution $(mm)(nn)$, and $ama'n \underset{\wedge}{\phantom{x}} bmb'n \underset{\wedge}{\phantom{x}} mbnb'$. We could not have $\langle am \rangle \equiv \langle bm \rangle$ without $a$ and $b$ coinciding. Hence $\langle am \rangle \equiv \langle mb \rangle$.

**9·54**   Corollary: *Angles that have a fixed pair of bisectors form an involution of line pairs.*

Another important property of angles is the following:

**9·55**   *If two angles $\langle ad \rangle$ and $\langle bc \rangle$ have the same bisectors, then*

$$\langle ab \rangle \equiv \langle cd \rangle.$$

[*] See Coolidge (Ref. 5, p. 29).

*Proof:* If *m* and *n* are the bisectors, the four pairs *ad, bc, a'd', b'c'* belong to the involution *(mm)(nn)*. Therefore $aba'b' \overline{\wedge} dcd'c' \overline{\wedge} cdc'd'$. Moreover, since $S(aba') \neq S(dcd')$, we have

$$S(aba') = S(cdd') = S(cdc').$$

In dealing with angles at different points, it is convenient to use another notation. If *a* is *AO* and *b* is *OB* (so that *A* and *B* are arbitrary points on the lines *a* and *b*, which intersect in *O*), we write ⟨*AOB*⟩ instead of ⟨*ab*⟩.

It should be emphasized that our definition of angle, while agreeing with Johnson's directed angle and Picken's cross (Sec. 1·7), differs from the customary definition, where the angle *AOB* would be changed into its supplement by moving *B* along the line *b* to the other side of the vertex *O*, and where *BOA* would be considered either equal to *AOB* or its negative (certainly not its supplement, as in the new treatment). The present convention has some definite advantages, chiefly in avoiding the separate consideration of various cases, as in the following theorem, where we should ordinarily have to make different statements according as S(*ACB*) agrees or disagrees with S(*ADB*):

**9·56**　*For any four points A, B, C, D on a circle,* ⟨*ACB*⟩ ≡ ⟨*ADB*⟩.

*Proof:* The angle ⟨*ACB*⟩ is measured by means of the four lines joining *C* to the vertices of the rectangle *ABA'B'* whose diagonals are diameters *AA'* and *BB'* of the circle, as in Fig. 9·5B. Joining *D*, instead of

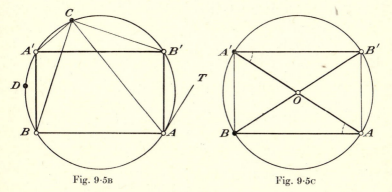

Fig. 9·5B　　　　　　Fig. 9·5c

*C*, to these same vertices, we obtain a related set of four lines and thence a congruent angle ⟨*ADB*⟩. The question of sense presents no difficulty, as the sense of the lines in either case agrees with S(*ABA'*) on the circle.

Conversely, the circle *ABC* could be described as the locus of a point *X* such that ⟨*ACB*⟩ ≡ ⟨*AXB*⟩.

Another famous theorem is Euclid I. 5 (*pons asinorum*):

**9·57**   *If* $OA \equiv OB$, *then* $\langle OAB \rangle \equiv \langle ABO \rangle$.

*Proof:* Let $AOA'$ and $BOB'$ be diameters of the circle with radii $OA$ and $OB$, as in Fig. 9·5c. Applying the half-turn about $O$ and using 9·56, we have

$$\langle OAB \rangle \equiv \langle OA'B' \rangle = \langle AA'B' \rangle \equiv \langle ABB' \rangle = \langle ABO \rangle*$$

Theorem 9·53 helps us to obtain the following less trivial result:

**9·58**   *The altitudes and sides of an oblique triangle bisect the angles of its pedal triangle.*

*Proof:* Let $ABC$ be the pedal triangle of $PQR$, as in Fig. 9·5D. Since

$ABC$ is the diagonal triangle of the quadrangle $PQRS$, the four lines drawn through $A$ (or $B$ or $C$) are a harmonic set, by 2·52.

Fig. 9·5D

It follows that each of the four points $P$, $Q$, $R$, $S$ lies in a different one of the four regions determined in the projective plane by the lines $BC$, $CA$, $AB$. If $S$ is the one lying in the finite region $ABC/o$, we say that $AS$ and $AQ$ are the *internal* and *external* bisectors of the angle $A$ of triangle $ABC$.

**9·59**   *The bisectors of the angles of any triangle concur in sets of three to form a quadrangle.*†

*Proof:* Let the external bisectors of the angles $B$ and $C$ of triangle $ABC$ meet in $P$. We merely have to construct, as in 2·42, the quadrangle $PQRS$ whose diagonal triangle is $ABC$.

Of the four points $P$, $Q$, $R$, $S$, the one in $ABC/o$ is called the *incenter* of triangle $ABC$, while the other three are the *excenters*.

### EXERCISES

1. If $AT$ is the tangent at $A$ to the circle $ABC$, prove that $\langle ACB \rangle \equiv \langle TAB \rangle$ (Fig. 9·5B).

2. Show that the center of a circle lies on a bisector of the angle formed by two intersecting tangents.

3. Prove that the incenter and excenters are the centers of four circles touching the sides of the triangle.

---

\* Here we are using the sign $\equiv$ for congruent angles and $=$ for identical angles. In Sec. 1·7, where angles were regarded as magnitudes, this distinction was unnecessary.

† von Staudt (Ref. 40, p. 209, §347).

4. Prove that the axes of a hyperbola bisect the angles between the asymptotes.

5. Prove that the asymptotes of a rectangular hyperbola bisect the angles between any pair of conjugate diameters.*

**9·6 Congruent Transformations.** We define the *reflection* in a line $m$ to be the harmonic homology whose axis is $m$ while its center is at infinity in the perpendicular direction. Thus any two points $A$ and $B$ are interchanged by the reflection in the right bisector of the segment $AB$, and any two intersecting lines $a$ and $b$ are interchanged by the reflection in either of the bisectors of the angle $\langle ab \rangle$. Moreover, if $OA$ and $OB$ are congruent segments on $a$ and $b$, a definite one of the angle bisectors will serve to reflect $A$ into $B$. By the dual of 5·27, the product of two harmonic homologies having the same center is an elation. Hence:

**9·61**　*The product of reflections in two parallel lines is a translation.*

We define a *rotation* about a point $O$ to be the product of reflections in two lines through $O$. One instance has already been considered: By 5·32 and Sec. 8·5:

**9·62**　*The product of reflections in two perpendicular lines is a half-turn.*

We define a *congruent transformation* to be a collineation that preserves length (and consequently preserves the line at infinity and the absolute involution). As instances we have a reflection, a translation, a rotation, and more generally the product of any number of reflections.

By 5·12, a collineation is determined by its effect on any quadrilateral. Hence a congruent transformation is determined by its effect on a triangle (which provides a quadrilateral when we add the line at infinity).

**9·63**　*A congruent transformation that leaves two points invariant is either the identity or a reflection.*

*Proof:* Let $P$ and $Q$ be the invariant points. Then the transformation relates two triangles $PQR$ and $PQR'$. If $R = R'$, we have the identity (by 5·11). Otherwise the altitudes from $R$ and $R'$ must have the same "foot" $C$, and $RC \equiv R'C$; thus the transformation is a harmonic homology with axis $PQ$, in fact, the reflection in $PQ$.

**9·64**　*Any congruent transformation that leaves just one point invariant is a rotation.*

* Chasles (Ref. 4, p. 117).

*Proof:* Let the given transformation take a triangle $PQR$ to $PQ'R'$. Then $PQ$ is transformed into $PQ'$ by the reflection in a definite one of the bisectors of $\langle QPQ' \rangle$, say $m$. We now have two triangles $PQ'R_1$ and $PQ'R'$, as in Fig. 9·6A. These must be distinct, since otherwise the

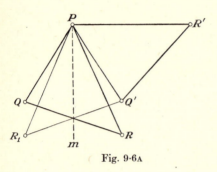

reflection in $m$ would suffice and every point on $m$ would be invariant. Hence the given transformation is the product of reflections in $m$ and $PQ'$.

Our next theorem is reminiscent of 4·22:

**9·65** *Any congruent transformation can be constructed as a product of reflections, the number of which can be reduced to three.*

Fig. 9·6A

*Proof:* If $PQR$ is transformed into an entirely distinct triangle $P'Q'R'$, we begin by reflecting in the right bisector of $PP'$ and then use one or two further reflections as above.

In the projective plane, the senses of two pencils may be compared only if the pencils have the same center. But in the affine plane the translation from the one center to the other enables us to compare the senses of any two pencils (see the remark at the end of Sec. 8·2). Thus we may classify congruent transformations as being *direct* or *opposite*, according to the relation between the senses of corresponding pencils. Since a reflection reverses the sense of every pencil, a congruent transformation is direct or opposite according as it is the product of an even or odd number of reflections. It follows from 9·63 and 9·64 that if an opposite transformation has an invariant point it must be a reflection. Therefore the product of reflections in any three concurrent lines is a reflection. Thus if $\Psi$ is a rotation about $O$, and $\Phi$ the reflection in any line through $O$, the product $\Phi\Psi$ is another reflection, say $\Phi'$, and

$$\Psi = \Phi\Phi\Psi = \Phi\Phi'$$

Hence:

**9·66** *Any rotation is the product of two reflections, one of which may be the reflection in any given line through the center of rotation.*

In particular, the half-turn about $O$ is the product of reflections in any two perpendicular lines through $O$.

We can now prove the converse of 9·55:

**9·67**  *If four concurrent lines satisfy* $\langle ab \rangle \equiv \langle cd \rangle$, *then the angles* $\langle ad \rangle$ *and* $\langle bc \rangle$ *have the same bisectors.*

*Proof:* Let $\Psi$ be the rotation that takes $a$ to $c$, and consequently $b$ to $d$, let $\Phi$ be the reflection in one of the bisectors of $\langle bc \rangle$, and let $\Phi'$ be the reflection $\Phi\Psi$. Then

$$\Phi = \Phi\Phi'\Phi' = \Phi\Phi\Psi\Phi\Psi = \Psi\Phi\Psi,$$

and this transforms $a$ into

$$a^{\Phi} = a^{\Psi\Phi\Psi} = c^{\Phi\Psi} = b^{\Psi} = d$$

Thus the given bisector of $\langle bc \rangle$ bisects $\langle ad \rangle$, too.

Interchanging $b$ and $c$ in 9·55, we deduce that the relation $\langle ab \rangle \equiv \langle cd \rangle$ implies $\langle ac \rangle \equiv \langle bd \rangle$. Hence:

**9·68**  *A rotation displaces different lines through congruent angles.*

We are now ready for an interesting specialization of Steiner's construction:

**9·69**  *The locus of the point of intersection of corresponding lines of two congruent pencils is a circle or a rectangular hyperbola according as the congruence is direct or opposite.* *

*Proof:* If the two pencils are directly congruent, they are related by a rotation and thus the angle between corresponding lines is constant and the locus is a circle. On the other hand, if the two pencils are oppositely congruent, there are two pairs of corresponding lines that are parallel (given by the invariant points of the hyperbolic projectivity induced on the line at infinity). By 6·54, the locus is a conic having asymptotes in these two directions. But when corresponding lines are parallel, the respectively perpendicular lines are likewise parallel; hence the two pairs mentioned above are perpendicular, and the locus is a rectangular hyperbola.

We now possess all the material needed for Euclid's development of congruent triangles and similar triangles; *e.g.*, two triangles are *similar* if there is a third triangle that is homothetic to the first and congruent to the second, or vice versa. Moreover, the theory of rotation leads to the *measurement* of angles, just as the theory of translation leads to the measurement of distance (Sec. 8·4). In fact, rotations about a point induce projectivities (resembling translations†) on the line at infinity.

---

* von Staudt (Ref. 40, p. 204, §337).

† For this development, see Coxeter (Ref. 6, Chap. V). For further results on congruent transformations, see Coxeter (Ref. 7, Chap. III).

## EXERCISES

1. Show that our definition of a congruent transformation is redundant, since any point-to-point correspondence that preserves length also preserves collinearity.

2. Show that the product of reflections in two parallel lines is a translation through twice the distance between the lines.

3. Deduce from 9·66 that any product of three reflections may be regarded as the product of a reflection and a half-turn. (It is interesting to compare this with a *rotation*, which is the product of two reflections, and a *translation*, which is the product of two half-turns.)

4. By a second application of 9·66, show that any product of three reflections may be regarded as a *glide reflection:* the product of a reflection and a translation that commute, the translation being along the axis of the reflection.

**9·7  Foci.**  Let us now return to the theory of conics and define foci by means of a property noticed by La Hire.* A *focus* of a conic is a point at which the involution of conjugate lines coincides with the involution of perpendicular lines. For instance, the circle has a focus at its center. (It remains to be seen whether such a point exists in any other case.) Since the involution of perpendicular lines through a point is elliptic:

**9·71**  *A focus is an interior point.*

(Any tangent that might be drawn through it, being self-conjugate, would have to be self-perpendicular.)

If the center is a focus, the conic must be a circle. If not, let $O$ be the center and $F$ a focus. Then $OF$ is an axis; for the chord through $F$ perpendicular to $OF$, being conjugate to $OF$, is bisected by $OF$.

**9·72**  *If there are two foci, their join is an axis.*

*Proof:* The lines through two foci $F$ and $F'$, perpendicular to their join $FF'$, are both conjugate to $FF'$. Therefore the pole of $FF'$ is at infinity, that is, $FF'$ is a diameter and, consequently, an axis.

It follows that any foci which exist must all lie on one axis. In the case of a hyperbola this must, by 9·71, be the transverse axis. To establish the existence of foci (one for a parabola, one for a circle, two for a hyperbola, and two for an eccentric ellipse), we shall describe a construction for them, making use of 6·57 (the dual of Seydewitz's theorem) as applied to a triangle $TT'U$ or $a'ab$, whose vertices $T$ and $T'$ are joined to $C$ (on $AA'$) by conjugate lines $p$ and $p'$, as in Fig. 9·7A.

---

* For the present treatment, see von Staudt (Ref. 40, p. 208, §345) and Reye (Ref. 31, p. 155).

In the case of a parabola, we take $a'$ to be the line at infinity, $a$ the tangent at the vertex $A$, and $b$ an arbitrary tangent (as in Fig. 9·7B), so that any point $C$ on the axis $AA'$ is joined to $T = a \cdot b$ and $T' = a' \cdot b$

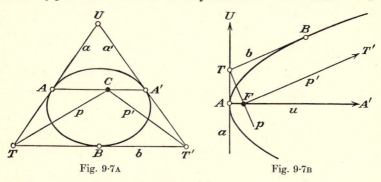

Fig. 9·7A                                       Fig. 9·7B

by conjugate lines. Then $C$ will be a focus if these lines are perpendicular; hence we have the following construction:

**9·73**   *To construct the focus $F$ of a parabola, let any tangent $b$ meet the tangent at the vertex in $T$. Through $T$ draw $p$ perpendicular to $b$. Then $p$ meets the axis in $F$.*

There is only one focus, for *any* focus would be joined to $T$ by a line perpendicular to $b$.

In the case of a hyperbola, we take $b$ to be an asymptote, while $a$ and $a'$ are the tangents at the two vertices $A$ and $A'$ (as in Fig. 9·7C), so that any point $C$ on the transverse axis $AA'$ is joined to $T = a \cdot b$

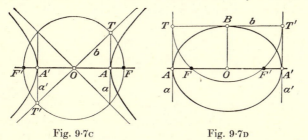

Fig. 9·7C                                       Fig. 9·7D

and $T' = a' \cdot b$ by conjugate lines. Then $C$ will be a focus if these lines are perpendicular. Hence:[*]

**9·74**   *To construct the foci $F$ and $F'$ of a hyperbola, let either asymptote meet the tangent at either vertex in $T$. Then the concentric circle through $T$ meets the transverse axis in $F$ and $F'$.*

[*] Holgate (Ref. 21, p. 220).

In the case of an ellipse, we take $b$ to be the tangent at one end of the minor axis (or any other tangent not parallel to the minor axis), while $a$ and $a'$ are the tangents at the ends of the major axis (as in Fig. 9·7D), so that any point $C$ on the major axis is joined to $T = a \cdot b$ and $T' = a' \cdot b$ by conjugate lines. Then $C$ will be a focus if these lines are perpendicular. Hence:*

**9·75**   *To construct the foci $F$ and $F'$ of an eccentric ellipse, let either of the tangents parallel to the major axis meet the two tangents parallel to the minor axis in $T$ and $T'$. Then the circle on $TT'$ as diameter meets the major axis in $F$ and $F'$.*

To make sure that this circle will meet the major axis, we observe that its radius $BT = OA$ is greater than $BO$; therefore the center $O$ of the ellipse is interior to the circle, and $AA'$ is a secant. If we interchanged the roles of the major and minor axes, we should have a circle of radius $OB$ that would fail to meet a line distant $OA$ from its center. Thus there are no foci on the minor axis (in real geometry).

### *EXERCISE*

Let $c$ be a variable line through a fixed point and $C$ its pole wo a given polarity. Prove that the line through $C$ perpendicular to $c$ envelops a parabola.

Fig. 9·8A

**9·8   Directrices.**   The polar of a focus is called a *directrix*. Before considering the general conic, let us establish two interesting properties of the directrix of a parabola.

**9·81**   *Perpendicular tangents to a parabola meet on the directrix.*

*Proof:* Let $a$ and $a'$ be the tangents to a parabola from a point $U$ on the directrix $f$, and let these meet the tangent at the vertex in $T$ and $T'$ (Fig. 9·8A). Since the focus $F$ is conjugate to $U$, 6·57 shows that $FT$ and $FT'$ are conjugate; therefore they are perpendicular. But $FT$ is perpendicular to $a$, and $FT'$ to $a'$. Hence $FTUT'$ is a rectangle, and $a$ is perpendicular to $a'$. Since any other tangent perpendicular to $a$ would be parallel to $a'$ (which is impossible), this completes the proof.

---

* Apollonius (Ref. 1, p. 424, lib. III, prop. 45).

**9·82**   *The orthocenter of a triangle circumscribed to a parabola lies on the directrix.**

*First Proof:* By the dual of 6·73, applied to the quadrangle formed by the sides of the given triangle and the line at infinity, the tangents to the parabola from the orthocenter $S$ are a line pair belonging to the same involution as the lines joining $S$ to the pairs of opposite vertices of this quadrilateral (which are the vertices of the triangle and the points at infinity on the opposite sides). Since these pairs of lines are perpendicular, so are the tangents.

*Second Proof:* Let $abc$ be the triangle and $o$ the line at infinity. Applying Brianchon's theorem (our 7·22) to the hexagon $abcc'oa'$, where $c'$ and $a'$ are the tangents perpendicular to $c$ and $a$, we conclude that the three lines

$$(a \cdot b)(c' \cdot o), \qquad (b \cdot c)(o \cdot a'), \qquad (c \cdot c')(a \cdot a')$$

are concurrent. The first two are altitudes of the triangle, and the third is the directrix of the parabola.

Fig. 9·8B

We come now to Pappus's celebrated construction for a conic as the locus of a point whose distance from a fixed point $F$ is $e$ times its distance from a fixed line $f$, where $e$ is a positive number called the *eccentricity*.

To establish this construction, let $F$ be a focus, $f$ its polar, $A$ the vertex between $F$ and $f$, $B$ any other point on the conic, $T$ the pole of $t = AB$, and $P = f \cdot t$ the pole of $FT$, as in Fig. 9·8B.† Now, $FT$ and $FP$, through

---

* This theorem is due to Steiner, the first proof to Holgate (Ref. 21, p. 209), and the second to J. C. Moore (see Salmon, Ref. 37, p. 247). It has seemed worth while to include both, because they are so neat.

† Reye (Ref. 31, pp. 159–160, slightly simplified).

the focus $F$, are conjugate and therefore perpendicular; also, they are harmonic conjugates wo $FA$ and $FB$ (by 6·41, applied to the conjugate points $FT \cdot t$ and $P$ on $AB$). Hence, by 9·53, they bisect the angle $\langle AFB \rangle$. The circle through $F$ with center $B$ will meet $FP$ again in a point $C$ such that $BC \equiv BF$, whence, by 9·57,

$$\langle BCF \rangle \equiv \langle CFB \rangle \equiv \langle AFC \rangle, \quad \text{that is,} \quad \langle BCP \rangle \equiv \langle AFP \rangle$$

which shows that $BC$ is parallel to the axis $AF$. Hence, if $A_1$ and $B_1$ are the points where $AF$ and $BC$ meet $f$,

$$\frac{CB}{BB_1} = \frac{FA}{AA_1}$$

Calling this ratio $e$, we have $FB = CB = eBB_1$.

### EXERCISES

1. Prove that the sum (or difference) of the two focal distances of a variable point on an ellipse (or hyperbola) is constant.

2. Prove that $e < 1$ for an ellipse, $e = 1$ for a parabola, $e > 1$ for a hyperbola, and $e = \sqrt{2}$ for a rectangular hyperbola.

3. Show that the conics which have a given point for focus and a given line for corresponding directrix form a self-dual system.

4. Prove that the orthocenters of the four triangles occurring in a quadrilateral are collinear.

# CHAPTER 10

# CONTINUITY

The purpose of this chapter is to show how, in the presence of the axioms of incidence and order, one very simple statement about limits will suffice for the derivation of all the celebrated properties of the one-dimensional continuum, including the axioms of Archimedes and Dedekind, and Enriques' theorem (our 3·51). This treatment may be regarded as the geometrical counterpart of Weierstrass' theory of irrational numbers.

**10·1   An Improved Axiom of Continuity.**   Our development of real projective geometry began with five axioms of incidence (2·21 to 2·25, or 2·31 to 2·35) and six axioms of order (3·11 to 3·16). These were sufficient to establish many interesting theorems, such as 3·21 and 3·33. But before we could prove that a projectivity (as defined by von Staudt) is an ordered correspondence, we had to introduce a twelfth axiom, 3·51, of a decidedly complicated nature. The corresponding statements in other books are no simpler, except when algebra is used and a direct appeal is made to the system of real numbers. The following purely geometrical axiom 10·11 is so simple that it requires only eight words. However, two of the words, *monotonic* and *limit*, must be carefully defined.

A sequence of points $A_0$, $A_1$, $A_2$, . . . is said to be *monotonic* if $A_0A_n//A_1A_{n+1}$ for every integer $n > 1$. (This just means that we have infinitely many points, arranged in cyclic order on the line.) The existence of such a sequence is ensured by Axiom 3·11.

A point $M$ is called a *limit* of this sequence $\{A_n\}$ if it satisfies the following two conditions:

(1) For every integer $n > 2$, $A_1A_n//A_2M$.

(2) For every point $P$ with $A_1P//A_2M$, there exists an $n$ such that $A_1A_n//PM$.

The above definitions are complete, but the following remarks (along with Fig. 10·1A) will perhaps help to clarify them. Regarding

$A_0$ as a kind of barrier, let us say "$X$ precedes $Y$" and write $X < Y$, if

$$S(A_0XY) = S(A_0A_1A_2)$$

and again if $Y = A_0$. (The transitivity of this relation is a consequence of our axioms of order.) Then the sequence $\{A_n\}$ is monotonic if

$$A_0 \quad A_1 \quad A_2 \qquad\qquad P \quad A_n \quad M$$

Fig. 10·1A

$A_1 < A_2 < \cdots$ ; and the two requirements for a limit $M$ are as follows:

(1) The points $A_1$, $A_2$, . . . all precede $M$.

(2) Every point that precedes $M$ precedes some $A_n$.

We are now ready for the axiom:

**10·11**  *Every monotonic sequence of points has a limit.*

We see at once that this limit is unique. For if $M$ and $M'$ are two such points, let $M'$ precede $M$. By (2), $M'$ precedes some $A_n$; but by (1) every $A_n$ precedes $M'$. Thus the assumption $M' < M$ leads to a contradiction; similarly, so does the assumption $M < M'$.

**10·2  Proving Archimedes' Axiom.**  In order to identify the real projective line with the one-dimensional continuum described by Cantor, we must examine various properties of the continuum and see whether we can deduce them from our axioms. For instance, the property of *density*, or *internal convexity*, is a consequence of Axiom 3·11. Defining *segment* as in Sec. 3·2, we may express this property as follows:

**10·21**  *Every segment contains a point.*

It follows that any segment contains infinitely many points.

A subtler property is what Hilbert[*] called the axiom of Archimedes (though it might more properly be ascribed to Eudoxus). He expressed it in affine terms, as follows: Let $A_1$ be any point between $A$ and $P$. Take points $A_1$, $A_2$, . . . so that

$$AA_1 \equiv A_1A_2 \equiv A_2A_3 \equiv \cdots$$

Then there exists a positive integer $n$ such that $P$ lies between $A$ and $A_n$.

Referring to Fig. 8·4A, we see that

$$H(MA_1, AA_2), \ H(MA_2, A_1A_3), \ \cdots$$

[*] Ref. 19, p. 25.

whence, by 3·21, $MA_1//AA_2$, $MA_2//A_1A_3$, $\ldots$ , *i.e.*,

$$\text{S}(MAA_1) = \text{S}(MA_1A_2) = \text{S}(MA_2A_3) = \cdots$$

Thus the sequence $\{A_n\}$ is monotonic, the projectivity

$$MAA_1X \overset{R}{\barwedge} MBB_1Y \overset{S}{\barwedge} MA_1A_2X'$$

(Fig. 10·2A) is direct, and $\text{S}(MAX) = \text{S}(MA_1X')$. Using $BY$ instead of $RS$, we have similarly

$$\text{S}(MAA_1) = \text{S}(MXX')$$

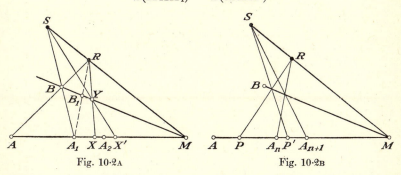

Fig. 10·2A          Fig. 10·2B

By 10·11, the monotonic sequence $\{A_n\}$ has a limit. The axiom of Archimedes asserts that this limit is precisely $M$. In other words:

**10·22**   *If* $\text{H}(MA_n, A_{n-1}A_{n+1})$ *for all positive integers n, then M is the limit of the sequence* $\{A_n\}$.

*Proof:* If the limit is not $M$, we can identify the monotonic sequence $MAA_1 \ldots$ with the $A_1A_2A_3 \ldots$ of Sec. 10·1. Then the limit must be some point $P'$ such that $MA_n//AP'$. We naturally express this relation as $A_n \prec P' \prec M$. Since $\text{S}(MAA_1) = \text{S}(MXX')$, we have also $X \prec X'$. Now construct the point $P'S \cdot MB$, which projects from $R$ into $P$ (so that the projectivity $X \barwedge X'$ would take $P$ to $P'$, as in Fig. 10·2B). Since $P$ precedes the limit $P'$, there must be an $A_n$ between $P$ and $P'$, and a consequent $A_{n+1}$ between $P'$ and $M$. Thus $P'$ precedes $A_{n+1}$, contradicting our assumption that $P'$ is the limit of the $A$'s. Hence there cannot in fact be a limit different from $M$.

**10·3   Proving the Line to Be Perfect.\*** The points $A, A_1, A_2, \ldots$ of 10·22 form what is sometimes called a *harmonic sequence*. This is part of Möbius's *harmonic net*, or *net of rationality*, which may be described as the smallest set of points that contains, for every three of

---

\* Russell (Ref. 35, p. 103; Ref. 36, p. 291). A range is said to be *perfect* if it satisfies 10·11, 10·21, and 10·31.

its members, the harmonic conjugate of each wo the other two. Any three points on the line lead to a harmonic net by repeated harmonic constructions, and it is easily seen* that the same harmonic net is equally well determined by any three of its points.

Cantor's continuum is not merely dense (in the sense of 10·21) and closed (in the sense of 10·11) but also "dense in itself": every point is the limit of a sequence. In particular:

**10·31**    *Each point is the limit of some monotonic sequence of points belonging to a given harmonic net.*

*Proof:* It is a corollary of 10·22 that any point of the given harmonic net is the limit of such a sequence; for we can construct the harmonic sequence $\{A_n\}$ of 10·22 from the three points $M$, $A$, $A_1$ that determine the harmonic net. Accordingly, let us take a point *not* belonging to the harmonic net and try to exhibit it as a limit.

Changing the notation slightly, let the given harmonic net be defined by three points $A$, $B$, $C$, and let $Z$ be a point not belonging to this net. By 3·14 and 3·17, $Z$ must occur in just one of the segments $BC/A$, $CA/B$, $AB/C$, say the last. For the sake of verbal economy, let us employ the language of affine geometry, regarding $C$ as the point at infinity on the line; *e.g.*, instead of "the harmonic conjugate of $C$ wo $A$ and $B$" we say simply "the mid-point of $AB$." This mid-point belongs to the net and decomposes the segment $AB$ (meaning $AB/C$ )into two parts, one of which must contain $Z$. That part is similarly decomposed by its mid-point. We continue indefinitely in this manner, always bisecting the part that contains $Z$, so as to obtain a *contracting sequence of segments*, each containing the next and all containing $Z$. We name such a segment† $LU$ in the order that makes $AU//LB$, so that $S(LUC) = S(ABC)$, which we write conventionally as $L \prec U$. The lower ends $L$ and upper ends $U$ form monotonic sequences of points, which, by 10·11, have limits $M$ and $N$ such that

$$L \prec M, \quad N \prec U$$

Since $L \prec Z \prec U$, just one of the following four statements must hold:

$$M = Z \prec N, \quad M \prec Z \prec N, \quad M \prec Z = N, \quad M = Z = N$$

We shall establish the last of these four by showing that the assumption $M \prec N$ leads to a contradiction.

---

\* Veblen and Young (Ref. 42, p. 85).

† We purposely avoid the notation $L_n U_n$, because any two consecutive segments have one end in common, and we are interested in sequences of *distinct* points.

Assuming $M \prec N$, let us construct points $M'$ and $N'$ so that

$$\text{H}(CN, MM') \qquad \text{and} \qquad \text{H}(CM, NN')$$

*i.e.*, so that $N$ is the mid-point of $MM'$ and $M$ of $NN'$. (This means, in effect, that the four points $N'$, $M$, $N$, $M'$ are "evenly spaced," as in Fig. 10·3A.) Since $M$ and $N$ are the limits of lower and upper ends, we

Fig. 10·3A

can find a lower end in the segment $N'M$, and an upper end in $NM'$. If these do not belong to the same one of the contracting sequence of segments, choose the latter one of the two segments involved. We thus obtain a segment $LU$ such that

$$N' \prec L \prec M \prec N \prec U \prec M'$$

Now construct points $L'$, $V$, $U'$ so that

$$\text{H}(CM, LL'), \qquad \text{H}(CV, LU), \qquad \text{H}(CN, UU')$$

*i.e.*, so that $M$, $V$, $N$ are the respective mid-points of $LL'$, $LU$, $UU'$. By 3·41 (applied to $M$ and $C$, with $X$ running from $L$ down to $N'$, as in Fig. 10·3B), we have $L' \prec N$; therefore $L' \prec U$. By 3·43 (applied to

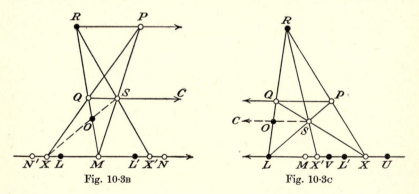

Fig. 10·3B          Fig. 10·3c

$L$ and $C$, with $X$ running from $L'$ up to $U$, as in Fig. 10·3c), $M \prec V$. Similarly, using $UU'$ instead of $LL'$, we find $V \prec N$. Thus $M \prec V \prec N$. But in the contracting sequence of segments, the next after $LU$ must be either $LV$, with $N \prec V$, or $VU$, with $V \prec M$. In either case we obtain a contradiction. Hence, instead of $M \prec N$, we must have $M = N$, and $Z$ is in fact the limit of both sequences $\{L\}$ and $\{U\}$.

We see now that the line is dense (containing a point between any two points), closed (containing the limit of each sequence) and dense in itself (each point a limit). According to Cantor, still one more property is needed before we can be sure that this continuum is strictly similar to the class of real numbers. The final requirement is the occurrence of an enumerable "relatively dense" subset.* A set is said to be *enumerable* if its members can be put into one-to-one correspondence with the integers (or with the rational numbers). A set of points is said to be *relatively dense* on the line if every segment contains a point of the set; in other words, there is not merely a point of the line between any two points of the special set (which follows from 10·21) but more surprisingly a point of the set between any two points of the line. This is the geometrical counterpart of the arithmetical theorem that a *rational* number can be found between any two *real* numbers.

Such an enumerable separation set is provided (for the real projective line) by a harmonic net. From the nature of its construction, this is obviously enumerable. It is relatively dense since, by 10·21 and 10·31:

**10·32** *Every segment contains a point of a given harmonic net.*

This is known as the Lüroth-Zeuthen theorem† because Lüroth and Zeuthen proved it independently in 1873, using a method which resembles our proof of 10·31. However, they assumed not only 10·11 (and 3·41 and 3·43) but also Dedekind's axiom, which we shall prove in Sec. 10·5.

**10·4  The Fundamental Theorem of Projective Geometry.**  The essential steps in our proof of the fundamental theorem 4·21 were 3·51, 3·62, 4·11, 4·12. How will the procedure be altered when continuity is given by 10·11 instead of 3·51? The simplest way is to use Pieri's definition for a segment (Sec. 3·6) so as to obtain 4·11 without any appeal to continuity. Then 4·12 will follow with the help of one simple lemma:

**10·41**  *If an ordered correspondence relates $A_n$ to $A'_n$, where $\{A_n\}$ is a monotonic sequence with limit $M$, then $\{A'_n\}$ is a monotonic sequence with limit $M'$.*

*Proof:* Since $A_1 \prec A_2 \cdots \prec M$ in the sense $S(A_0 A_1 A_2)$, we must have

$$A'_1 \prec A'_2 \cdots \prec M'$$

in $S(A'_0 A'_1 A'_2)$. Hence $\{A'_n\}$ is monotonic, and its limit $N'$ cannot follow $M'$ in the latter sense. On the other hand, if $N'$ preceded $M'$, it would

---

* Forder (Ref. 12, p. 14). This is called a *median class* in Russell (Ref. 35, p. 104).

† See Whitehead (Ref. 44, pp. 30–33), or Mathews (Ref. 25, pp. 43–46).

come from a point $N$ separated from $M$ by some $A_n$, implying

$$N' \prec A'_n \prec M'$$

whereas every $A'_n$ should precede the limit $N'$. Hence in fact $N'$ must coincide with $M'$.

To prove 4·12, we recall that a projectivity is a correspondence that preserves the harmonic relation. Thus, if three points are invariant, the whole harmonic net determined by those three points must be invariant. Hence, by 4·11, 10·41, and 10·31, *every* point is invariant.

**10·5  Proving Dedekind's Axiom.** Dedekind's axiom may be expressed as follows:

**10·51**  *For every division of all the points of a given segment or interval $\alpha$ into two nonempty sets $R_1$ and $R_2$, such that every point of $R_1$ precedes every point of $R_2$, there exists a point $M$ in $\alpha$ which has the property that every point of $\alpha$ preceding $M$ belongs to $R_1$ and every point of $\alpha$ following $M$ belongs to $R_2$.*

In other words, if $L \prec U$ for every point $L$ of the lower set and every point $U$ of the upper, then one of the two sets contains a dividing point $M$ such that $L \prec M \prec U$, except that either $L$ or $U$ might coincide with $M$.

The proof closely resembles that of 10·31. Let $\alpha$ be $AB/C$ or $\overline{AB}/C$, and consider its mid-point, *i.e.*, the harmonic conjugate of $C$ wo $A$ and $B$. Call the mid-point $L$ or $U$ according as it belongs to $R_1$ or $R_2$. Similarly bisect $LB$ or $AU$, as the case may be. In this manner we obtain a contracting sequence of segments, each having one end in $R_1$ and the other in $R_2$. We see, as before, that the lower ends and upper ends have the same limit $M$, which is Dedekind's dividing point.

**10·6  Enriques' Theorem.** We are now ready to prove the theorem that we regarded as an axiom in 3·51. We take first the case when the correspondence is direct:

**10·61**  *If a direct correspondence relates an interval $\overline{AB}/C$ to an interior interval $\overline{A'B'}/C$, then the latter contains an invariant point $M$ such that there is no invariant point between $A$ and $M$ (in $\overline{AB}/C$).*

*Proof:* If $A$ is invariant, there is no more to be said: $M$ coincides with $A$. If not, suppose $A$ is related to another point $A'$, $A'$ to $A''$, and so on. Then the iterated correspondence provides a monotonic sequence $AA'A''\ \dots$, whose limit $M$ is invariant by 10·41 (applied to the sequences $AA'A''\ \dots$ and $A'A''A'''\ \dots$). A point between $A$ and $M$ cannot be invariant. For if it coincides with some $A^{(n)}$, it is related to the different point $A^{(n+1)}$; and if it lies in a segment $A^{(n-1)}A^{(n)}$, its corresponding point lies in the different segment $A^{(n)}A^{(n+1)}$.

This is fairly simple. The difficult case (because iteration cannot help) is when the correspondence is opposite:

**10·62** *If an opposite correspondence relates an interval $\overline{AB}/C$ to an interior interval $\overline{A'B'}/C$, then the latter contains an invariant point.*

*Proof:*[*] Assuming that there is no invariant point, we derive a contradiction by the following argument: We use Dedekind's axiom 10·51, taking the lower set to consist of those points of $\overline{AB}/C$ which *precede* their corresponding points, while the upper set consists of those which *follow* their corresponding points. These sets are easily seen to have the requisite properties. First, the sets are not empty; for since $B' \prec A'$, the lower contains $A$ and the upper $B$. Second, if $L$ precedes its corresponding point $L'$ while $U$ follows $U'$, $L$ must precede $U$; for otherwise we should have $U' \prec U \prec L \prec L'$, so that $LU$ and $L'U'$ would have the same sense. By our assumption the dividing point $M$ is not invariant but related to a distinct point $M'$.

Since $AL$ and $A'L'$ have opposite senses, any point $L$ of the lower set satisfies $A \prec L \prec L' \prec A'$. Thus $A'$, following every point of the lower set, must belong to the upper set. Similarly $B'$ belongs to the lower set.

Now consider a fixed point $L$ between $A$ and $M$, and a variable point $X$ between $L$ and $M$, so that $A \prec L \prec X \prec M$. Since $X$ belongs to the lower set and the correspondence is opposite, we have $X \prec X' \prec L'$. Thus $L'$ follows *every* point $X$ that precedes $M$, and hence either follows or coincides with $M$.

If $M'$ precedes $M$ (so that $A \prec M' \prec M \prec A'$, or possibly $A \prec M' \prec M = A'$), then every point $L'$ between $M$ and $M'$, being also between $A'$ and $M'$, comes from some point $L$ between $A$ and $M$, whereas we have just seen that any such $L$ yields an $L'$ which does *not* precede $M$. Hence $M'$ cannot precede $M$.

Similarly, by supposing $M \prec Y \prec U \prec B$ for a fixed $U$ and variable $Y$, we find that $U'$ either precedes or coincides with $M$, and deduce that $M'$ cannot follow $M$.

Thus, finally, $M' = M$, contradicting our assumption that no point is invariant. Therefore some point must be invariant (and it follows, as in Sec. 3·5, that the invariant point is unique).

[*] Cf. Enriques (Ref. 11, pp. 71–75).

# CHAPTER 11

# THE INTRODUCTION OF COORDINATES

In Chap. 10 we discussed many properties of the real projective line. But there remain certain questions that would be difficult, if not impossible, to answer without using the concept of a coordinate or abscissa. For instance, how can you be sure that a harmonic net does not exhaust all the points on the line?

We saw, in Sec. 8·4, that pairs of points on the affine line belong to an involution if their algebraic distances from a fixed point on the line have either a constant sum or a constant product. The question now arises: What is the projective counterpart of this affine statement? More precisely: What projective entities can be added or multiplied? One answer was given by von Staudt,* who used sets of four points, which he called *Würfe* (*i.e.*, "throws" or "casts"). Hessenberg, in 1905, simplified that treatment by fixing three of the four points and operating with the remaining one. Instead of adding or multiplying segments $OX$, as in the affine line, we now add or multiply points $X$ in the presence of three fixed points, which play the role of the numbers 0, 1, ∞. (Anyone familiar with the vectorial approach to analytic geometry will understand how a vector $OX$ and its end point $X$ are for many purposes interchangeable.)

Following O'Hara and Ward,† we develop this theory "in one dimension," using elementary properties of involutions (instead of constructions involving arbitrary points outside the given line). Although this method obscures Hilbert's famous discovery of the connection between Pappus's theorem and the commutativity of multiplication, it has the advantage of allowing the range of points to be on a conic just as well as on a line. We shall see, in Figs. 11·1A and 11·2A, how very easily the sum or product of two points on a conic may be constructed.‡

* Ref. 41, pp. 166–194.
† Ref. 26, pp. 155–156. Cf. Veblen and Young (Ref. 42, pp. 141–156).
‡ Ref. 42, p. 232.

The chief novelty arises in Sec. 11·8, where we introduce two-dimensional homogeneous coordinates. The use of a conic makes it unnecessary to mention either cross ratio or nonhomogeneous coordinates.

**11·1  Addition of Points.**  Relative to two fixed points $P_0$ and $P_\infty$ on a given line or conic, we define the *sum* $A + B$ of two arbitrary

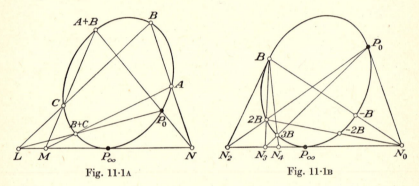

Fig. 11·1A                    Fig. 11·1B

points (on the same line or conic, but distinct from $P_\infty$) to be the companion of $P_0$ in the hyperbolic involution

$$(AB)(P_\infty P_\infty)$$

Thus $A + P_0 = A$, and $A + A$ is the harmonic conjugate of $P_0$ wo $A$ and $P_\infty$. The commutative law

$$A + B = B + A$$

is satisfied immediately, and the equation $X + B = A$ can be solved by taking $X$ to be the companion of $B$ in $(AP_0)(P_\infty P_\infty)$. In particular, $-B$ (such that $-B + B = P_0$) is the harmonic conjugate of $B$ wo $P_0$ and $P_\infty$.

To establish the associative law

$$A + (B + C) = (A + B) + C$$

we observe that $P_\infty$ is an invariant point of both the involutions

$$(A B)(A + B P_0)    \text{and}    (C B)(B + C P_0)$$

whence, by 4·73, $P_\infty$ is also an invariant point of $(A B + C)(A + B C)$. Thus $A$ and $B + C$ have the same sum as $A + B$ and $C$.*

We now define $2B = B + B$, $3B = 2B + B$, and so on. Since

* This method is due to Alex Rosenberg.

$$(m - 1)B + (m + 1)B = mB + mB$$

each successive multiple $(m + 1)B$ arises as the harmonic conjugate of $(m - 1)B$ wo $mB$ and $P_\infty$.

The operation of adding $B$ can be expressed as a projectivity. In fact, the above definitions imply

$$P_\infty P_0(-B)A \barwedge P_\infty P_0 B(-A) \barwedge P_\infty B P_0(A + B)$$

If $B$ is a fixed point (not $P_0$ or $P_\infty$), this combined projectivity is independent of the choice of $A$, and hence it relates a variable point $X$ to $X + B$. This still holds when $X$ is $P_\infty$, provided that we extend the definition of addition* by declaring that, if $B \neq P_\infty$, $P_\infty + B = P_\infty$.

By 3·21 and 3·32, the projectivity

**11·11**                    $P_\infty P_0(-B) \barwedge P_\infty B P_0$

is direct. (In fact, since $P_\infty$ is the only invariant point, it is parabolic.) Applying it repeatedly to any point (other than $P_\infty$), we obtain a monotonic sequence. In particular:

**11·12**   *The sequence of points*

$$\ldots \; -3B, \; -2B, \; -B, \; P_0, \; B, \; 2B, \; 3B, \; \ldots$$

*is monotonic.*

In other words, the relation $S(kB \; mB \; P_\infty) = S(P_0 \; B \; P_\infty)$ holds if and only if $k < m$.

### EXERCISES

1. Derive 11·12 from the fact that $mB$ and $P_\infty$ separate $(m \pm 1)B$.

2. Show that the sum of two points $A$ and $B$ on a conic may be constructed as in Fig. 11·1A: $AB$ meets the tangent at $P_\infty$ in $N$ (the center of the additive involution), and $A + B$ is the point where the line $NP_0$ meets the conic again. Deduce that $P_\infty + B = P_\infty$ $(B \neq P_\infty)$.

3. Derive the associative law for addition from Pascal's theorem (our 7·21) applied to the hexagon $A \; B \; C \; (A + B) \; P_0 \; (B + C)$.

4. Let $P_0 B$ (Fig. 11·1A or B) meet the tangent at $P_\infty$ in $N_1$. Show that the point of contact of the second tangent from $N_1$ is a point $\frac{1}{2}B$ such that $\frac{1}{2}B + \frac{1}{2}B = B$.

---

* This convention can be justified by appealing to the degenerate involution

$$(P_\infty B)(P_\infty P_\infty),$$

which relates every point (in particular, $P_0$) to $P_\infty$.

5. Given the parabola in its familiar Cartesian form $y = x^2$, prove that the point on it with abscissa $a + b$ may be located by drawing through the vertex $(0, 0)$ the chord parallel to $(a, a^2)(b, b^2)$, as in Fig. 11·1c. Show how this agrees with the formal addition of points on the parabola.

Fig. 11·1c

**11·2  Multiplication of Points.** Relative to three fixed points $P_0$, $P_1$, $P_\infty$, on a given line or conic, we define the *product* $A \times B$ of two arbitrary points (on the same line or conic, but distinct from $P_0$ and $P_\infty$) to be the companion of $P_1$ in the elliptic or hyperbolic involution

$$(AB)(P_0 P_\infty)$$

For example, $A \times P_1 = A$. The commutative law, $A \times B = B \times A$, is satisfied immediately, and the equation $X \times B = A$ can be solved by taking $X$ to be the companion of $B$ in $(AP_1)(P_0 P_\infty)$. In particular, $B^{-1}$ (such that $B^{-1} \times B = P_1$) is the companion of $B$ in the hyperbolic involution $(P_1 P_1)(P_0 P_\infty)$, whose second invariant point is $-P_1$, or say $P_{-1}$; thus $B^{-1}$ is the harmonic conjugate of $B$ wo $P_1$ and $P_{-1}$.

To establish the associative law

**11·21**    $$A \times (B \times C) = (A \times B) \times C$$

we observe that $P_0 P_\infty$ is a pair of both the involutions

$$(A \, B)(A \times B \, P_1) \quad \text{and} \quad (C \, B)(B \times C \, P_1)$$

whence, by 4·68, $P_0 P_\infty$ is also a pair of $(A \, B \times C)(A \times B \, C)$. Thus $A$ and $B \times C$ have the same product as $A \times B$ and $C$.

Since $A^{-1} \times B^{-1} \times B \times A = P_1$, we have

$$A^{-1} \times B^{-1} = (B \times A)^{-1} = (A \times B)^{-1}$$

The operation of multiplying by $B$ can be expressed as a projectivity. By 2·71 and the definition of $A \times B$,

$$P_0 P_\infty P_1 A \; \overline{\wedge} \; P_\infty P_0 A P_1 \; \overline{\wedge} \; P_0 P_\infty B (A \times B)$$

If $B$ is a fixed point (not $P_0$ or $P_\infty$), this combined projectivity is independent of the choice of $A$ and hence it relates a variable point $X$ to $X \times B$. This still holds when $X$ is $P_0$ or $P_\infty$, provided that we extend the definition of multiplication* by declaring that, if $B \neq P_\infty$,

---

\* These conventions can be justified by means of the degenerate involutions

$$(P_0 B)(P_0 P_\infty) \quad \text{and} \quad (P_\infty B)(P_\infty P_0).$$

$P_0 \times B = P_0$ and, if $B \neq P_0$, $P_\infty \times B = P_\infty$.

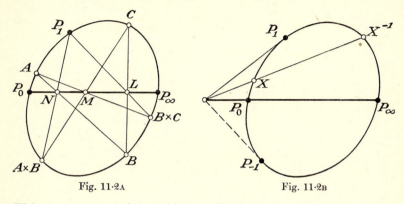

Fig. 11·2A                                 Fig. 11·2B

This suggests an alternative proof for the associative law 11·21. Using the projectivity

**11·22**                        $P_0 P_\infty P_1 \barwedge P_0 P_\infty C$

which relates every $X$ to $X \times C$, we have

$$P_0 P_\infty (A \times B) A \barwedge P_\infty P_0 P_1 B \barwedge P_\infty P_0 C(B \times C)$$

Thus $A$ and $B \times C$ have the same product as $A \times B$ and $C$.

Similarly, to prove the distributive law

**11·23**              $(A \times C) + (B \times C) = (A + B) \times C$

we observe that, from the definition of $A + B$,

$$P_\infty A B P_0 \barwedge P_\infty B A (A + B)$$

whence, by 11·22,

$$P_\infty (A \times C)(B \times C) P_0 \barwedge P_\infty (B \times C)(A \times C)[(A + B) \times C]$$

This involution exhibits $(A + B) \times C$ as the sum of $A \times C$ and $B \times C$.

By repeated application of 11·23 we see that, for any positive integer $n$, $n(A \times C) = (nA) \times C$. In particular,

$$nC = P_n \times C, \qquad \text{where } P_n = nP_1$$

If $m$ is the greater of two positive integers $m$ and $n$, we have

$$-mC + nC + (m - n)C = -mC + mC = P_0$$

whence $-mC + nC = -(m - n)C$; similarly

$$-mC + (-nC) = -(m + n)C.$$

Accordingly, we may regard $(-n)C$ as having the same meaning as $-(nC)$.

1. Verify that $(-A) \times (-B) = A \times B$. (*Hint:* Use Sec. 4·6, Exercise 3.)

2. Show that the product of two points $A$ and $B$ on a conic may be constructed as in Fig. 11·2A; $AB$ meets $P_0 P_\infty$ in $N$ (the center of the multiplicative involution), and $A \times B$ is the point where the line $NP_1$ meets the conic again. Deduce that $P_0 \times B = P_0$ ($B \neq P_\infty$) and $P_\infty \times B = P_\infty$ ($B \neq P_0$).

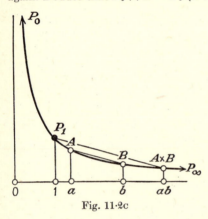

Fig. 11·2c

3. Derive the associative law for multiplication from Pascal's theorem (our 7·21), applied to the hexagon $A\,B\,C\,(A \times B)\,P_1\,(B \times C)$.

4. Verify that the projectivity 11·22 is direct if $P_0$ and $P_\infty$ do not separate $P_1$ and $C$ (*e.g.*, if these points are arranged as in Fig. 11·2A). In this case $P_0 P_\infty$ meets $P_1 C$ in an exterior point, from which we can draw two tangents to the conic. Show that their points of contact are $\pm C^{\frac{1}{2}}$, such that $C^{\frac{1}{2}} \times C^{\frac{1}{2}} = C$.

5. Given the rectangular hyperbola $xy = 1$ or $y = x^{-1}$, prove that the point with abscissa $ab$ may be located by drawing through the vertex $(1, 1)$ the chord parallel to $(a, a^{-1})(b, b^{-1})$, as in Fig. 11·2c. Show how this agrees with the multiplication of points on the rectangular hyperbola.

**11·3  Rational Points.**  Defining

$$P_n = nP_1, \quad P_{-n} = -P_n, \quad P_{1/n} = (P_n)^{-1}, \quad P_{m/n} = mP_{1/n}, \quad P_{-m/n} = -P_{m/n}$$

we obtain a definite point $P_a$ for every rational number $a$. We call $P_a$ a *rational point* and $a$ its *abscissa*.

The addition and multiplication of such points agree with the addition and multiplication of the corresponding numbers. For if $a = k/n$ and $b = m/n$ (where $k$ and $m$ are integers, while the common denominator $n$ is a positive integer), we have:

$$P_a + P_b = kP_{1/n} + mP_{1/n} = (k + m)P_{1/n} = P_{(k+m)/n} = P_{a+b}$$

Again, if $a = k/l$ and $b = m/n$ (where $l$ and $n$ are positive integers),

$$P_a \times P_b = P_k \times P_{1/l} \times P_m \times P_{1/n} = P_{km} \times P_{1/l} \times P_{1/n} = kmP_{1/ln}$$
$$= P_{km/ln} = P_{ab}$$

Moreover, the *order* of the points $P_a$ agrees with the order of the rational numbers $a$. For to see whether $P_a$ precedes or follows $P_b$ we express $a$ and $b$ in terms of a common denominator, say $a = k/n$, $b = m/n$, and observe where $P_a$ and $P_b$ occur in the sequence 11·12 with $B = P_{1/n}$. We conclude that

$$S(P_aP_bP_\infty) = S(P_0P_1P_\infty)$$

if and only if $a < b$.

**11·4　Projectivities.** Setting $B = P_b$ in 11·11, we obtain a projectivity that relates a variable point $P_x$ to $P_{x+b}$. Thus the transformation of abscissas

$$x' = x + b$$

represents a projectivity $P_x \barwedge P_{x+b}$, which is parabolic if $b \neq 0$.

Similarly, setting $C = P_a$ in 11·22, we obtain a projectivity that relates $P_x$ to $P_{ax}$. Thus the transformation

$$x' = ax \qquad (a \neq 0)$$

represents a projectivity $P_x \barwedge P_{ax}$, which is hyperbolic if $a \neq 1$.

The product of these two elementary transformations is

**11·41**
$$x' = ax + b \qquad (a \neq 0)$$

The third kind of elementary transformation

$$x' = \frac{1}{x}$$

represents the hyperbolic involution $(P_0P_\infty)(P_1P_1)$, as in Fig. 11·2B.

By judiciously combining all three elementary transformations, we obtain the linear fractional transformation

$$x' = \frac{a}{c} + \frac{b - ad/c}{cx + d} = \frac{ax + b}{cx + d}$$

where $c \neq 0$ and $b \neq ad/c$. These inequalities can be weakened to

$$ad \neq bc$$

for by setting $c = 0$ (and $d = 1$) we obtain 11·41. Since $P_\infty^{-1} = P_0$, we can take care of the possibility that $x = \infty$ by writing the linear fractional transformation in the two alternative forms

**11·42**
$$x' = \frac{ax + b}{cx + d} = \frac{a + b/x}{c + d/x} \qquad (ad - bc \neq 0)$$

Another way of writing it is

$$cxx' - ax + dx' - b = 0$$

The projectivity thus represented is an involution if $x$ and $x'$ are inter-changeable, *i.e.*, if $-a = d$. In particular, the involution with invariant points $P_a$ and $P_b$ is represented by

$$xx' - \tfrac{1}{2}(a + b)(x + x') + ab = 0$$

Hence:

**11·43**   *The relation* $H(P_aP_b, P_cP_d)$ *is equivalent to*

$$(a + b)(c + d) = 2(ab + cd)$$

### EXERCISES

1. Show that the parabolic projectivity 11·11, as applied to points on a conic, has the tangent at $P_\infty$ for its axis.

2. Show that the hyperbolic projectivity 11·22 (with $C \neq P_1$) has the secant $P_0P_\infty$ for its axis.

3. Show that the projectivity 11·41 is hyperbolic except when $a = 1$, and that it is direct or opposite according as $a$ is positive or negative.

**11·5   The One-dimensional Continuum.**   The various steps by which we have derived the general rational point $P_a$ from three arbi-trary points $P_0$, $P_1$, $P_\infty$ may all be expressed in terms of harmonic conjugacy:

$$H(P_1P_\infty, P_0P_2), \qquad H(P_mP_\infty, P_{m-1}P_{m+1}), \qquad H(P_0P_\infty, P_mP_{-m}),$$
$$H(P_1P_{-1}, P_nP_{1/n}), \qquad H(P_{m/n}P_\infty, P_{(m-1)/n}P_{(m+1)/n}), \qquad H(P_0P_\infty, P_aP_{-a})$$

Conversely, by 11·43, the harmonic conjugate of $P_c$ wo $P_a$ and $P_b$ is another rational point $P_d$. Hence:

**11·51**   *The rational points* $P_a$, *along with* $P_\infty$, *form a harmonic net.*

(For this reason, a harmonic net is sometimes called a *net of ration-ality*.)

We are now ready to show how the remaining points of the range may be included in this algebraic treatment by defining irrational abscissas.

Let any real number $x$ be expressed as the limit of a monotonic sequence of rational numbers $a$. By Axiom 10·11 (which naturally holds on the conic just as well as on the line), the corresponding sequence of points $P_a$ has a definite limit, which we denote by $P_x$.

Conversely, by 10·31, any given point on the line or conic may be regarded as the limit of a monotonic sequence of rational points. The corresponding sequence of rational numbers $a$ is eventually monotonic in the algebraic sense (after possibly discarding some initial terms of the wrong sign); thus it is either divergent or convergent. In the former case, the given point must have been $P_\infty$; in the latter, the limit of the $a$'s is a real number $x$, and the point is $P_x$.

The number $x$, whether real or infinite, is called the *abscissa* of $P_x$. Thus, when the three fundamental points $P_0, P_1, P_\infty$ have been assigned, every point on the line or conic has a uniquely determined abscissa.

Since the number of real numbers is strictly greater than the number of rational numbers,* we see at last why it is that the harmonic net constitutes only an "infinitesimal part" of the whole range.

We can now remove the restriction to rational abscissas in Sec. 11·4. With real numbers $a$, $b$, $c$, $d$, 11·42 is the most general projectivity; for† $P_0P_1P_\infty \overline{\wedge} P_pP_qP_r$ is given by

$$x' = \frac{r(p - q)x + p(q - r)}{(p - q)x + (q - r)}$$

which is a valid transformation provided that $p$, $q$, $r$ are all different. Hence, if a variable point $P_x$ (on a line or a conic) is projectively related to $P_{x'}$ (on the same line or conic or another), then the abscissas $x$ and $x'$ must be connected by a linear fractional transformation. In particular, the most general projectivity preserving $P_\infty$ is 11·41 (which takes $P_0$ to $P_b$ and $P_1$ to $P_{a+b}$).

Since $P_0P_1P_\infty P_x \overline{\wedge} P_0P_aP_\infty P_{ax}$, any construction by which $P_x$ is derived from $P_0P_1P_\infty$ will yield $P_{ax}$ when applied to $P_0P_aP_\infty$. Thus $P_x$ can be renamed $P_{ax}$ provided $P_1$ is renamed $P_a$. (The new $P_1$ is the old $P_{1/a}$.) More generally, instead of regarding the transformation 11·42 as a projectivity, we may equally well regard it as a consistent renaming of all the points (without altering any of their geometrical properties). Such renaming is called a *change of scale*.

We have seen (Sec. 11·4) that the general involution is

$$cxx' - a(x + x') - b = 0 \qquad (a^2 + bc \neq 0)$$

This reduces to $x + x' = k$ when $c = 0$, and to

$$(x - a)(x' - a) + g = 0 \qquad (g \neq 0)$$

otherwise. By a simple change of scale (namely, $x \to x + \frac{1}{2}k$ or $x + a$) these relations become

**11·52**　　　　　　　　　　$x + x' = 0$

and $xx' + g = 0$. The former, having invariant points $P_0$ and $P_\infty$, may be taken as the canonical form for a hyperbolic involution. The latter

---

* Russell (Ref. 35, pp. 85–86).

† Veblen and Young (Ref. 42, p. 155).

is elliptic if $g > 0$, in which case we can make a further simplification by the change of scale $x \to x \sqrt{g}$. Thus the canonical form for an elliptic involution is:

**11·53**                         $$xx' + 1 = 0$$

### EXERCISES

1. The harmonic net based on three collinear points $A$, $B$, $C$ naturally contains $D$, the harmonic conjugate of $C$ wo $A$ and $B$. Show that it does *not* contain the invariant points of the hyperbolic involution $(AD)(BC)$. *Hint:* Take the abscissas of $A$, $B$, $C$, $D$ to be $\infty$, 1, 2, 0.

2. If to the harmonic net we adjoin such further points arising from every set of four points already obtained, have we then exhausted all the points on the line?

3. Show how Sec. 7·5, Exercise 5, would enable us to express any elliptic involution in the form $(P_0 P_\infty)(P_1 P_{-1})$, thereby justifying 11·53 immediately.

4. With any two points on a line (or conic) we may associate the quadratic equation whose roots are their abscissas. If three point pairs form a quadrangular set (Sec. 4·7), prove that their equations

$$p_i x^2 + q_i x + r_i = 0 \qquad (i = 1, 2, 3)$$

satisfy

$$\begin{vmatrix} p_1 & q_1 & r_1 \\ p_2 & q_2 & r_2 \\ p_3 & q_3 & r_3 \end{vmatrix} = 0$$

*Hint:* If the point pairs belong to the involution

$$b + a(x + x') - cxx' = 0$$

we have

$$bp_i - aq_i - cr_i = 0$$

**11·6  Homogeneous Coordinates.**  Given any five points of which no three are collinear, we can draw a definite conic through them, take an arbitrary sixth point $P_1$ on the conic, and name two of the given points $P_0$ and $P_\infty$. Then the remaining three points have definite abscissas $x_1$, $x_2$, $x_3$; and a different choice of the sixth point $P_1$ would have the effect of multiplying all of $x_1$, $x_2$, $x_3$ by some number $a$.

We call $x_1$, $x_2$, $x_3$ the *coordinates* of the point $P_\infty$ for the *triangle of reference* $P_{x_1} P_{x_2} P_{x_3}$ and *unit point* $P_0$. From the above remarks we see that these coordinates are quite definite, apart from the possibility of multiplying all three by the same constant; in other words, they are *homogeneous* coordinates. We denote $P_\infty$ by the symbol $(x_1, x_2, x_3)$, with the understanding that for any nonzero $a$, the symbol $(ax_1, ax_2, ax_3)$ denotes the same point.

The above definition implies that the numbers $x_1$, $x_2$, $x_3$ are all distinct and different from zero. We shall see later that this restriction can be removed, except that they must not be all zero. In particular, the name *unit point* will be justified when we have found the coordinates 1, 1, 1 for $P_0$.

It should be noticed that to find the coordinates for a new point, we generally have to start over again with a new conic. Moreover, we cannot yet construct a point with given coordinates.

**11·7  Proof that a Line Has a Linear Equation.**  By projecting the points of a conic from a fixed point (on the conic) onto a line, we obtain points on the line, which may be regarded as having the same abscissas, referred to the projections of the fundamental points $P_0$, $P_1$, $P_\infty$. In particular, by projecting the points $P_0$, $P_\infty$, $P_{x_1}$, $P_{x_2}$, $P_{x_3}$ of Sec. 11·6 from $P_{x_2}$ onto the line $P_{x_1}P_{x_3}$ and from $P_{x_1}$ onto $P_{x_2}P_{x_3}$, we obtain points whose abscissas on those sides of the triangle (left and right) are as indicated in Fig. 11·7A. Applying the changes of scale

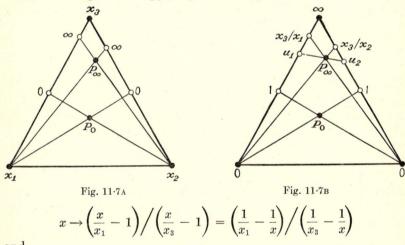

Fig. 11·7A          Fig. 11·7B

$$x \to \left(\frac{x}{x_1} - 1\right)\Big/\left(\frac{x}{x_3} - 1\right) = \left(\frac{1}{x_1} - \frac{1}{x}\right)\Big/\left(\frac{1}{x_3} - \frac{1}{x}\right)$$

and

$$x \to \left(\frac{x}{x_2} - 1\right)\Big/\left(\frac{x}{x_3} - 1\right) = \left(\frac{1}{x_2} - \frac{1}{x}\right)\Big/\left(\frac{1}{x_3} - \frac{1}{x}\right)$$

to the respective sides, we obtain the revised abscissas indicated in Fig. 11·7B.

Let any line through $P_\infty$ meet these sides in the points $u_1$ and $u_2$. Since

$$\infty \; 0 \frac{x_3}{x_1} u_1 \; \overline{\underset{\wedge}{\phantom{x}}}^{P_\infty} \; \infty \; \frac{x_3}{x_2} 0 \; u_2$$

the connection between $u_1$ and $u_2$ is of the form 11·41, *viz.*, since the values 0 and $x_3/x_1$ for $u_1$ correspond to the values $x_3/x_2$ and 0 for $u_2$,

**11·71**
$$x_1 u_1 + x_2 u_2 - x_3 = 0$$

We can fix this line by fixing the values of the two numbers $u_1$ and $u_2$, which determine definite points on those two sides of the triangle in terms of 0, 1, $\infty$. Now let the point $P_\infty$ vary on the line. This means that the numbers $x_1$, $x_2$, $x_3$ will vary; but they will continue to satisfy 11·71, which may thus be regarded as the *equation* of the line. This merely means that it is the condition for the variable point $(x_1, x_2, x_3)$ to lie on the line.

The point of intersection of two such lines is obtained by solving the two simultaneous equations for $x_1:x_2:x_3$. In this way we may extend the definition of coordinates to points on the sides of the triangle of reference or on the lines joining the vertices to the unit point $P_0$. Thus the point marked $u_2$ in Fig. 11·7B is where the line 11·71 meets another such line with a different $u_1$, namely

$$(0, 1, u_2)$$

*e.g.*, the point marked 1 on that same side is $(0, 1, 1)$. We now see that the characteristic property of a point on the first side of the triangle of reference is the vanishing of the first coordinate; hence the three sides have the equations

$$x_1 = 0, \qquad x_2 = 0, \qquad x_3 = 0$$

Consequently the vertices, where these sides meet in pairs, are

$$(1, 0, 0), \qquad (0, 1, 0), \qquad (0, 0, 1)$$

Setting $u_1 = 0$ in 11·71, we obtain the line $u_2 x_2 - x_3 = 0$, which joins $(1, 0, 0)$ to $(0, 1, u_2)$. This holds for all values of $u_2$ except, possibly, $u_2 = 1$; accordingly, we extend the meaning of coordinates so as to make it hold there, too. Thus the lines joining $P_0$ to the vertices of the triangle have the equations

$$x_2 = x_3, \qquad x_3 = x_1, \qquad x_1 = x_2$$

and the point $P_0$ itself is $(1, 1, 1)$.

We now see how to obtain coordinates for any given point. Conversely, given three real numbers $x_1$, $x_2$, $x_3$, not all zero, we can locate the point $(x_1, x_2, x_3)$ as follows: If two of the coordinates are zero, the point is a vertex. If one is zero, the point lies on a side; for example, $(0, x_2, x_3)$ has abscissa $x_3/x_2$ referred to

$$P_0 = (0, 1, 0), \qquad P_1 = (0, 1, 1), \qquad P_\infty = (0, 0, 1)$$

If none is zero, we join $(1, 0, 0)$ to $(0, x_2, x_3)$ and $(0, 1, 0)$ to $(x_1, 0, x_3)$, locating $(x_1, x_2, x_3)$ as the point where these joins intersect.

**11·8   Line Coordinates.**   Finally, we restore the symmetry of the three coordinates by writing 11·71 in the homogeneous form

$$x_1 X_1 + x_2 X_2 + x_3 X_3 = 0,$$

or

**11·81**                    $$X_1 x_1 + X_2 x_2 + X_3 x_3 = 0$$

and we call the coefficients $X_i$ the *coordinates* (line coordinates, envelope coordinates, or tangential coordinates) of the line

$$[X_1, X_2, X_3]$$

This device enables us to interchange points and lines in accordance with the principle of duality. The equation 11·81 is essentially self-dual, being the condition for the line $[X_1, X_2, X_3]$ and point $(x_1, x_2, x_3)$ to be incident. If we fix the point instead of the line, it is the condition for a variable line to pass through a fixed point; *i.e.*, the point $(x_1, x_2, x_3)$ has the equation

$$x_1 X_1 + x_2 X_2 + x_3 X_3 = 0$$

and its coordinates are the coefficients of $X_1$, $X_2$, $X_3$ in its equation. In particular, the vertices of the triangle of reference and the unit point have the equations

$$X_1 = 0, \qquad X_2 = 0, \qquad X_3 = 0, \qquad \text{and} \qquad X_1 + X_2 + X_3 = 0$$

In terms of line coordinates, the sides are

$$[1, 0, 0], \qquad [0, 1, 0], \qquad [0, 0, 1]$$

and the lines joining the vertices to the unit point are

$$[0, 1, -1], \qquad [-1, 0, 1], \qquad [1, -1, 0]$$

### EXERCISES

1. Show that the unit line $[1, 1, 1]$ meets the sides of the triangle of reference in the points $(0, 1, -1)$, $(-1, 0, 1)$, $(1, -1, 0)$.

2. Show that the two points $(0, 1, \pm x)$ are harmonic conjugates wo $(0, 1, 0)$ and $(0, 0, 1)$. Deduce the condition $x_1 X_1 = x_2 X_2 = x_3 X_3$ for the point $(x_1, x_2, x_3)$ and line $[X_1, X_2, X_3]$ to be trilinear pole and polar wo the triangle of reference.

3. Verify the result of Exercise 2 without using harmonic conjugates, by obtaining the coordinates of the various points and lines in Fig. 7·7A, beginning with $A = (1, 0, 0)$, $B = (0, 1, 0)$, $C = (0, 0, 1)$, and $X = (x_1, x_2, x_3)$.

# CHAPTER 12

# THE USE OF COORDINATES

In Chap. 11 we saw how a system of coordinates is inherent in synthetic geometry. In the present chapter we shall reverse the process, building up the analytic geometry from first principles, and deriving the theorems (including the axioms) from properties of numbers. We shall find the analytic method enables us to solve some problems more easily. On the other hand, it would be a grave mistake to abandon the synthetic method, which is far more stimulating to one's geometrical ingenuity.

**12·1  Consistency and Categoricalness.\***  In Chaps. 2 to 7 and 10 we developed the geometry of the real projective plane as a logical system based on the primitive concepts *point, line, incidence, separation,* and the twelve axioms 2·21 to 2·25, 3·11 to 3·16, and 10·11. This system has two essential properties: it is consistent and it is categorical. Before attempting to define these terms, let us remark that the properties are tested by means of *models,* wherein the primitive concepts, instead of remaining undefined, are defined in terms of concepts sufficiently familiar to be taken for granted. To establish the validity of a model, we merely have to verify that the given definitions for the primitive concepts enable us to *prove* the axioms.

When we say that a logical system is *consistent,* we mean that it cannot lead (by any chains of deduction, however long) to two contradictory propositions. The existence of a single model suffices to establish consistency; for any two contradictory results would imply contradictory properties of the model, and the absurdity would be manifest. The chief difficulty is to find an entirely satisfactory model.

Many would be prepared to take for granted, as a matter of experience, the ordinary geometry of Euclid or the affine geometry that can be extracted from it. Then we can define ideal elements as in Sec. 1·4

---

\* Cf. Veblen and Young (Ref. 42, pp. 1–6). The relation between synthetic and analytic geometry has been very ably described by Robson (Ref. 33, Chap. 8; Ref. 34, Chap. 19).

and verify that the affine plane, plus its points at infinity and line at infinity, forms a model for the projective plane.

Others might reject this model on the grounds that the space of experience is only approximately Euclidean. Setting aside the question as to whether a straight line is better approximated by a taut string or a ray of light, they would argue that Euclid's postulate of parallelism (which is an essential part of affine geometry) can be tested experimentally only in a neighborhood that is very small from the astronomical standpoint. Such persons might be prepared to take for granted the *local* properties of ordinary space. Then a model for the real projective plane is provided by the lines and planes through a fixed point in space. These lines and planes represent the points and lines of the projective plane, while incidence and order retain their customary meaning. This model has the great advantage of symmetry: there is no "line at infinity" to play a special role.

Still others might object that even this symmetrical model rests on intuitive ideas of space that cannot be justified by purely logical means. For them we must devise a model which every geometrical concept is defined in terms of numbers. The validity of such an analytic model will be verified in Secs. 12·3 and 12·4. Of course, there remains the question of the consistency of the number system, but at that stage the geometer delegates his responsibility to the algebraist.

When we say that a logical system is *categorical*, we mean that it is unique, in the sense that any model is isomorphic with any other. Thus a geometry is categorical if the entities which represent all the points and lines in one model can be put into correspondence (one-to-one) with those which represent the points and lines in another model. The results of Chap. 11 serve to establish the categoricalness of our system of real projective geometry. For they provide a definite naming of all the points and lines by sets of three real numbers, and this naming can be carried over into each model.

However, the whole problem of consistency and categoricalness is connected with very difficult and deep questions, which have lately been investigated by philosophers and logicians, notably Gödel. Any adequate discussion would be beyond the scope of this book.

*EXERCISE*

Consider the following model for the geometry defined by the axioms of incidence (2·21 to 2·25) alone. *Points* are the 13 symbols $A_0, A_1, \ldots, A_{12}$; *lines* are the 13 symbols $a_0, a_1, \ldots, a_{12}$; $A_i$ and $a_j$ are incident if

$$i - j \equiv 0, 1, 3 \text{ or } 9 \pmod{13}$$

Deduce that the "geometry of incidence" is not categorical (Veblen).

**12·2 Analytic Geometry.** We have remarked that the most satisfactory way to establish the logical consistency of our axioms, without taking any geometrical ideas for granted, is by means of an algebraic model. Such a model will now be described in detail.

A *point* is defined as an ordered set of three real numbers $(x_1, x_2, x_3)$ not all zero, with the understanding that $(\lambda x_1, \lambda x_2, \lambda x_3)$ is the same point for any nonzero $\lambda$. Likewise a *line* is an ordered set of three real numbers $[X_1, X_2, X_3]$, not all zero, with the understanding that $[\lambda X_1, \lambda X_2, \lambda X_3]$ is the same line. For brevity we speak of the point $(x)$ and the line $[X]$. This point and line are said to be *incident* (the point lying on the line and the line passing through the point) if and only if

**12·21** $$\{xX\} = 0$$

where

$$\{xX\} = x_1 X_1 + x_2 X_2 + x_3 X_3 = \Sigma x_i X_i$$

Any figure or argument can be dualized by interchanging small and capital letters, round and square brackets.

If $(x)$ is a variable point on a fixed line $[X]$, we call 12·21 the *equation* of the line $[X]$, meaning that it is the necessary and sufficient condition for $(x)$ to lie on $[X]$. Dually, if $[X]$ is a variable line through a fixed point $(x)$, we call the same relation the *equation* of the point $(x)$, meaning that is the condition for $[X]$ to pass through $(x)$. Thus the coordinates of a line or point are the coefficients in its equation (see Sec. 11·8).

The three points $(1, 0, 0)$, $(0, 1, 0)$, $(0, 0, 1)$ and the three lines $[1, 0, 0]$, $[0, 1, 0]$, $[0, 0, 1]$ form a triangle called the *triangle of reference* (see Fig. 12·2A). The point $(1, 1, 1)$ and line $[1, 1, 1]$ are called the *unit point* and *unit line*. We shall see, in Sec. 12·5, that there is nothing geometrically special about this triangle and point and line, apart from the fact that the point and line are trilinear pole and polar wo the triangle.

By eliminating $X_1 : X_2 : X_3$ from the three equations

$$\{xX\} = 0, \qquad \{yX\} = 0, \qquad \{zX\} = 0$$

we find that the necessary and sufficient condition for three points $(x)$, $(y)$, $(z)$ to be collinear is

**12·22**
$$\begin{vmatrix} x_1 & x_2 & x_3 \\ y_1 & y_2 & y_3 \\ z_1 & z_2 & z_3 \end{vmatrix} = 0$$

This condition is equivalent to the existence of numbers $\lambda$, $\mu$, $\nu$, not all zero, such that

$$\lambda x_i + \mu y_i + \nu z_i = 0 \qquad (i = 1, 2, 3)$$

If $(y)$ and $(z)$ are distinct points, $\lambda \neq 0$. Hence the general point collinear with $(y)$ and $(z)$ is $(\mu y_1 + \nu z_1, \mu y_2 + \nu z_2, \mu y_3 + \nu z_3)$ or, briefly,

$$(\mu y + \nu z)$$

When $\mu = 0$, this is the point $(z)$ itself. For any other position, we can

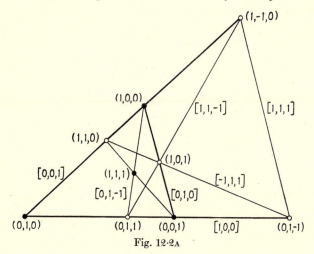

Fig. 12·2A

allow the coordinates of $(y)$ to absorb the $\mu$, and the collinear point is simply

$$(y + \nu z)$$

If we are concerned with only one such point, we may allow the $\nu$ to be absorbed too; thus three distinct collinear points may be expressed as $(y)$, $(z)$, $(y + z)$. However, this last simplification cannot be effected simultaneously on two lines if thereby one point would have to absorb two different parameters.

To illustrate these ideas, let us find the harmonic conjugate of $(y + z)$ wo $(y)$ and $(z)$. Referring to Fig. 2·5A, let the points

$$A, \qquad B, \qquad C, \qquad\qquad Q, \qquad R$$

be

$$(y), \qquad (z), \qquad (y + z), \qquad (x), \qquad (x + y)$$

Then $P$, on both $QC$ and $RB$, must be $(x + y + z)$; $S$, on both $QB$ and

$PA$, must be $(x + z)$; and $D$, on both $AB$ and $RS$, must be $(y - z)$. Hence, replacing $z_i$ by $vz_i$:

**12·23**    *The harmonic conjugate of $(y + vz)$ wo $(y)$ and $(z)$ is $(y - vz)$.*

Since this result is independent of $(x)$, we have here an analytic proof of 2·51.

Dually, the condition for three lines $[X]$, $[Y]$, $[Z]$ to be concurrent is

**12·24** $$\begin{vmatrix} X_1 & X_2 & X_3 \\ Y_1 & Y_2 & Y_3 \\ Z_1 & Z_2 & Z_3 \end{vmatrix} = 0$$

The general line concurrent with $[Y]$ and $[Z]$ is $[\mu Y + \nu Z]$, any such line except $[Z]$ can be expressed as

$$[Y + \nu Z]$$

and its harmonic conjugate wo $[Y]$ and $[Z]$ is $[Y - \nu Z]$.

### EXERCISES

1. Show that the line joining $(1, 0, 0)$ to $(x_1, x_2, x_3)$ is $[0, x_3, -x_2]$. Where does it meet $[1, 0, 0]$?

2. Name three lines through the point $(-1, 0, 1)$. Find their points of intersection with $[0, 0, 1]$.

3. If the triangle of reference is the diagonal triangle of a quadrangle having $(1, 1, 1)$ for one vertex, where are the other three vertices? (Cf. 2·42.)

4.* Show that the lines $[X]$ and $(y)(z)$ meet in the point

$$(\{zX\}y - \{yX\}z)$$

*Hint:* What is the condition for $(y + \nu z)$ to lie on $[X]$?

**12·3 Verifying the Axioms of Incidence.** To show that this analytic geometry really forms a model for the synthetic geometry developed in Chap. 2, we must verify that Axioms 2·21 to 2·25 are all satisfied.

The first four are easy. The point $(1, 0, 0)$ and line $[1, 0, 0]$ are certainly not incident. A line $[X_1, X_2, X_3]$ with $X_1 X_2 X_3 \neq 0$ is incident with three points such as $(0, X_3, -X_2)$, $(-X_3, 0, X_1)$, $(X_2, -X_1, 0)$; $[0, X_2, X_3]$ is incident with $(0, X_3, -X_2)$, $(1, X_3, -X_2)$, $(1, 0, 0)$; and so on. Two points $(y)$ and $(z)$ are incident with the unique line 12·22 or

$$\left[ \begin{vmatrix} y_2 & y_3 \\ z_2 & z_3 \end{vmatrix}, \begin{vmatrix} y_3 & y_1 \\ z_3 & z_1 \end{vmatrix}, \begin{vmatrix} y_1 & y_2 \\ z_1 & z_2 \end{vmatrix} \right]$$

---

* Graustein (Ref. 14, p. 70, Exercise 5). The idea of using capital letters for line coordinates is due to G. T. Bennett.

and two lines $[Y]$ and $[Z]$ are incident with the point 12·24 or

$$\left( \begin{vmatrix} Y_2 & Y_3 \\ Z_2 & Z_3 \end{vmatrix}, \begin{vmatrix} Y_3 & Y_1 \\ Z_3 & Z_1 \end{vmatrix}, \begin{vmatrix} Y_1 & Y_2 \\ Z_1 & Z_2 \end{vmatrix} \right)$$

As for Desargues' theorem (our 2·25), let $P$, $Q$, $R$, and the point of concurrence be $(x)$, $(y)$, $(z)$, and $(u)$. Then there is no loss of generality in taking $P'$, $Q'$, $R'$ to be

$$(x + u), \qquad (y + u), \qquad (z + u)$$

The point $QR \cdot Q'R'$, being collinear with $(y)$ and $(z)$ and also with $(y + u)$ and $(z + u)$, can only be $(y - z)$; similarly $RP \cdot R'P'$ and $PQ \cdot P'Q'$ are $(z - x)$ and $(x - y)$. The collinearity of these three points follows from the identity

$$(y - z) + (z - x) + (x - y) = 0$$

## EXERCISES

1. Find coordinates for the point of intersection of $[0, 1, -1]$ and $[-1, 1, 1]$ (Fig. 12·2A).

2. Prove Desargues' theorem as applied to the triangle of reference and $(k_1, 1, 1)(1, k_2, 1)(1, 1, k_3)$.

3. Work out Sec. 4·3, Exercise 3, taking $A_1A_2A_3$ to be the triangle of reference while $B_1B_3B_2$ is $(k_1, 1, 1)(1, k_2, 1)(1, 1, k_3)$. Obtain $k_1k_2k_3 = 1$ as the condition for lines $[0, -1, k_2]$, $[k_3, 0, -1]$, $[-1, k_1, 0]$ to be concurrent.

4. What further condition is required in Exercise 3 if also the lines $A_1B_1$, $A_2B_2$, $A_3B_3$ are concurrent (so that the two triangles are "quadruply perspective")?

**12·4   Verifying the Axioms of Order and Continuity.** To show that this analytic geometry suffices for a model of the real projective geometry of Chap. 3, we still have to verify that Axioms 3·11 to 3·16 and 10·11 are satisfied when separation is suitably defined. The definition we shall adopt is as follows. Denoting points $A$, $B$, $C$, $D$ by $(a)$, $(b)$, $(c)$, $(d)$, where

$$c_i = a_i + \mu b_i \qquad \text{and} \qquad d_i = a_i + \nu b_i$$

we say that $AB//CD$ if and only if

$$\frac{\nu}{\mu} < 0$$

(so that $\mu$ and $\nu$ have opposite signs). We saw, in Sec. 12·2, that any third point collinear with two distinct points $(a)$ and $(b)$ can be expressed as $(a + \nu b)$ where $\nu \neq 0$; hence the above expressions for $c_i$ and $d_i$ merely mean that $C$ and $D$ are collinear with $A$ and $B$.

Possibly the criterion $\nu/\mu < 0$ seems arbitrary; but it is really forced upon us by Pieri's definition for a segment (Sec. 3·6). According to that definition, the segment $(ACB)$ is the locus of the harmonic conjugate of $C$ wo two points $(a \pm \lambda b)$, where $\lambda$ takes all values except 0. Since $C$ is $(c)$ where

$$c_i = a_i + \mu b_i = \frac{(\lambda + \mu)(a_i + \lambda b_i) + (\lambda - \mu)(a_i - \lambda b_i)}{2\lambda}$$

its harmonic conjugate $(x)$ is given by

$$x_i = (\lambda + \mu)(a_i + \lambda b_i) - (\lambda - \mu)(a_i - \lambda b_i) = 2\mu \left( a_i + \frac{\lambda^2}{\mu} b_i \right)$$

*i.e.*, the harmonic conjugate is $\left( a + \dfrac{\lambda^2}{\mu} b \right)$, where $\lambda$ varies while $\mu$ is fixed. Since $\lambda$ is real, the coefficient $\lambda^2/\mu$ takes in turn every value having the *same* sign as $\mu$. Hence the supplementary segment $AC/B$ consists of all points $(a + \nu b)$ for which $\nu$ has the *opposite* sign.

Axiom 3·11 is verified by taking $\nu = -\mu$. The next two axioms are immediate. (Since $\nu/\mu < 0$, we cannot have $\mu = 0$ or $\nu = 0$ or $\mu = \nu$.) To test 3·14, we observe that

$$a_i = c_i - \mu b_i, \qquad d_i = c_i - (\mu - \nu)b_i$$

$$\mu b_i = c_i - a_i, \qquad \frac{\mu}{\nu} d_i = c_i - \left( 1 - \frac{\mu}{\nu} \right) a_i$$

Thus the three relations* $CB//AD$, $CA//BD$, $AB//CD$ mean

$$1 - \frac{\nu}{\mu} < 0, \qquad 1 - \frac{\mu}{\nu} < 0, \qquad \frac{\nu}{\mu} < 0$$

one of which must hold whenever $\mu\nu(\mu - \nu) \neq 0$.

As for 3·15, the relations $AB//CD$ and $CA//BE$ mean that $C, D, E$ are $(a + \mu b)$, $(a + \nu b)$, $(c + \rho a)$, where $\nu/\mu < 0$ and $-\rho < 0$. Since $E$ is $(a + \mu b + \rho a)$ or $\left( a + \dfrac{\mu}{1 + \rho} b \right)$, the relation $AB//DE$ means that

$$\frac{\mu}{(1 + \rho)\nu} < 0$$

which is obviously true if $\mu/\nu < 0$ and $\rho > 0$.

To test 3·16, suppose $ABCD \overset{S}{\doublebarwedge} A'B'C'D'$. We may take $S, A', B'$ to be $(s)$, $(a + \kappa s)$, $(b + \lambda s)$, and deduce

---

* Since $c_i/\mu = b_i + a_i/\mu$ and $d_i/\nu = b_i + a_i/\nu$, the relation $AB//CD$ is equivalent to $BA//CD$. This justifies our use of $CB//AD$ in place of $BC//AD$.

$$C' = A'B' \cdot CS = (a + \kappa s + \mu \overline{b + \lambda s})$$
$$D' = A'B' \cdot DS = (a + \kappa s + \nu \overline{b + \lambda s})$$

so that the relation $A'B'//C'D'$ means $\nu/\mu < 0$ again.

According to the definition in Sec. 10·1, a sequence of collinear points

$$A_0 = (z), \qquad A_1 = (y), \qquad A_n = (y + \nu_n z) \quad (n = 2, 3, \ldots)$$

is monotonic if $A_0A_n//A_1A_{n+1}$ (for every $n > 1$). Since

$$A_1 = \left(z - \frac{y + \nu_n z}{\nu_n}\right) \qquad \text{and} \qquad A_{n+1} = \left(z + \frac{y + \nu_n z}{\nu_{n+1} - \nu_n}\right)$$

this condition amounts to $\nu_n/(\nu_{n+1} - \nu_n) > 0$, or

$$\frac{\nu_{n+1}}{\nu_n} > 1 \qquad (n = 2, 3, \ldots)$$

which means that we have a monotonic sequence of numbers

$$\nu_1 = 0, \nu_2, \nu_3, \ldots$$

We know from analysis that such a sequence of numbers is either convergent (to some limit $\nu$) or divergent. In either case the sequence of points has a limit: $(y + \nu z)$ or $(z)$, respectively.

We have now completed the identification of our synthetic and analytic geometries. But a few further remarks arise naturally at this stage. Comparing the above results (*e.g.*, 12·23) with Sec. 11·5, we can identify the parameter $\nu$ with the abscissa of the point $(y + \nu z)$, referred to fundamental points

$$P_0 = (y), \qquad P_1 = (y + z), \qquad P_\infty = (z)$$

If $ABC$ and $A'B'C'$ are any two sets of three collinear points, we may write

$$A = (a), \qquad B = (b), \qquad C = (a + b)$$
$$A' = (a'), \qquad B' = (b'), \qquad C' = (a' + b')$$

Then the analytic verification of the fundamental theorem 4·21 consists in the observation that if $ABCD \; \overline{\wedge} \; A'B'C'D'$, the abscissas of $D$ and $D'$ agree:

$$D = (a + \nu b) \qquad \text{and} \qquad D' = (a' + \nu b')$$

This parameter $\nu$ is called the *cross ratio* of the four collinear points. In the notation of Veblen and Young it is $R(AB, DC)$.

More generally, if $A$, $B$, $C$, $D$ are

$$(a), \qquad (b), \qquad (a + \mu b), \qquad (a + \nu b),$$

we have $R(AB, DC) = \nu/\mu$, or $R(AB, CD) = \mu/\nu$. Thus the relation

$ABCD \mathrel{\overline{\wedge}} A'B'C'D'$ is equivalent to $R(AB, CD) = R(A'B', C'D')$,

$\quad AB//CD \qquad$ is equivalent to $R(AB, CD) < 0$,

and $\quad H(AB, CD) \quad$ is equivalent to $R(AB, CD) = -1$.

Let $[X]$ be any line through $C$ and $[Y]$ any line through $D$. Then

$$\{aX\} + \mu\{bX\} = 0, \qquad \{aY\} + \nu\{bY\} = 0$$

and therefore

$$\frac{\mu}{\nu} = \frac{\{aX\}\{bY\}}{\{bX\}\{aY\}}$$

Following Heffter and Koehler,[*] let us call this the *cross ratio* of the two points $(a)$, $(b)$ and the two lines $[X]$, $[Y]$. In particular, the two points are separated by the two lines if and only if the cross ratio is negative.

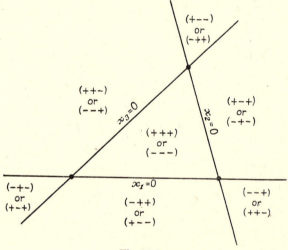

Fig. 12·4A

One simple consequence of this theory is the analytic interpretation of Theorem 3·81 as applied to the triangle of reference. If $(a)$ and $(b)$ are two points such that $a_1a_2b_1b_2 \neq 0$, their cross ratio with the sides $[1, 0, 0]$ and $[0, 1, 0]$ is

* Ref. 16, pp. 120, 136.

$$\frac{\{aX\}\{bY\}}{\{bX\}\{aY\}} = \frac{a_1 b_2}{b_1 a_2} = \frac{a_1}{a_2} \Big/ \frac{b_1}{b_2}$$

Hence the two angular regions bounded by the lines $x_1 = 0$ and $x_2 = 0$ are distinguished by the sign of $x_1/x_2$, or of $x_1 x_2$; and the four triangular regions determined by the three lines $x_i = 0$ are distinguished by the signs of $x_2 x_3$, $x_3 x_1$, $x_1 x_2$. In the "interior" region containing $(1, 1, 1)$ the three coordinates $x_i$ all have the same sign; but in the remaining three regions one coordinate differs in sign from the other two (see Fig. 12·4A).

A similar distinction can be made as to the signs of the coordinates of a line. If $\{xX\} = 0$, the three products $x_i X_i$ cannot all have the same sign. Hence any line for which $X_1 X_2 X_3 \neq 0$ must be exterior to the region where the point coordinates have the same distribution of signs as these line coordinates.

### EXERCISES

1. Show that the three relations $AB//CD$, $AB//CE$, $AB//DE$ cannot all hold simultaneously (cf. 3·18).

2. Prove in detail that if $\{\nu_n\}$ is a divergent monotonic sequence of numbers, the limit of the sequence of points $(y + \nu_n z)$ is $(z)$. *Hint:* Write $(y + \nu_n z)$ in the form $(z + \nu_n^{-1} y)$.

## 12·5  The General Collineation.  Consider the transformation

**12·51**
$$\begin{cases} x_i' = c_{i1} x_1 + c_{i2} x_2 + c_{i3} x_3 = \Sigma c_{ij} x_j & (i = 1, 2, 3) \\ X_j = c_{1j} X_1' + c_{2j} X_2' + c_{3j} X_3' = \Sigma c_{ij} X_i' & (j = 1, 2, 3) \end{cases}$$

where

$$\begin{vmatrix} c_{11} & c_{12} & c_{13} \\ c_{21} & c_{22} & c_{23} \\ c_{31} & c_{32} & c_{33} \end{vmatrix} = \Delta \neq 0$$

(The $\Sigma$ implies summation over the three values of $i$ or $j$, whichever of these letters appears *twice* in the expression.) This transformation leads from a point $(x)$ to a point $(x')$ and from a line $[X]$ to a line $[X']$. Since

$$\{x'X'\} = \Sigma x_i' X_i' = \Sigma\Sigma c_{ij} x_j X_i' = \Sigma x_j X_j = \{xX\}$$

it preserves incidence; and since $\Delta \neq 0$, it is one-to-one. Hence it is a collineation, as defined in Sec. 4·1.

Solving the equations 12·51 for $x_j$ and $X_i'$, we obtain the following expressions for $[X']$ in terms of $[X]$ and for $(x)$ in terms of $(x')$:

**12·52**
$$\begin{cases} X'_i = C_{i1}X_1 + C_{i2}X_2 + C_{i3}X_3 = \Sigma C_{ij}X_j & (i = 1, 2, 3) \\ x_j = C_{1j}x'_1 + C_{2j}x'_2 + C_{3j}x'_3 = \Sigma C_{ij}x'_i & (j = 1, 2, 3) \end{cases}$$

where $C_{ij}$ is the cofactor of $c_{ij}$ divided* by the determinant $\Delta$, so that

$$\Sigma c_{ij}C_{ik} = \delta_{jk}$$

which means 1 or 0 according as $j = k$ or $j \neq k$ (the "Kronecker delta"). These solutions may be verified as follows:

$$\Sigma C_{ij}X_j = \Sigma\Sigma C_{ij}c_{kj}X'_k = \Sigma\delta_{ik}X'_k = X'_i$$
$$\Sigma C_{ij}x'_i = \Sigma\Sigma C_{ij}c_{ik}x_k = \Sigma\delta_{jk}x_k = x_j$$

Given a triangle $(a)$ $(b)$ $(c)$, we may describe the position of any point $P$ by means of *barycentric* coordinates, defined as follows: If $P$ does not coincide with the vertex $(a)$, it can be joined to $(a)$ by a definite line that meets the opposite side $(b)(c)$ in a point $(\mu b + \nu c)$. Then $P$, being collinear with $(a)$ and $(\mu b + \nu c)$, may be expressed as

$$(\lambda a + \mu b + \nu c)$$

The barycentric coordinates are these coefficients $\lambda$, $\mu$, $\nu$. The point $(a)$ itself is included by allowing both $\mu$ and $\nu$ to vanish. By absorption we may take any particular point not on a side of the triangle to have $\lambda = \mu = \nu = 1$.

When $(a)(b)(c)$ and $(a + b + c)$ are the triangle of reference and unit point, $(\lambda a + \mu b + \nu c)$ is $(\lambda, \mu, \nu)$ and the barycentric coordinates are the same as the ordinary coordinates. The collineation

**12·53**
$$x'_i = a_i x_1 + b_i x_2 + c_i x_3$$

transforming the quadrangle $(1, 0, 0)(0, 1, 0)(0, 0, 1)(1, 1, 1)$ into $(a)(b)(c)(a + b + c)$, transforms $(\lambda, \mu, \nu)$ into the point

$$(\lambda a + \mu b + \nu c)$$

which has these same barycentric coordinates referred to the new quadrangle instead of the old.

Since *any* quadrangle can be written as $(a)(b)(c)(a + b + c)$, we see from 5·12 that 12·51 or 12·53 is the *most general* collineation. The condition $\Delta \neq 0$ or

$$\begin{vmatrix} a_1 & b_1 & c_1 \\ a_2 & b_2 & c_2 \\ a_3 & b_3 & c_3 \end{vmatrix} \neq 0$$

---

* Since the coordinates are homogeneous, there would be no harm in defining $C_{ij}$ to be just the cofactor of $c_{ij}$, without troubling to divide by $\Delta$.

merely expresses the requirement that the points $(a)$, $(b)$, $(c)$ form a triangle. Thus the equations 12·51 may be regarded *either* as a collineation shifting the points in the plane *or* as a coordinate transformation giving a new name to each point.

Defining

$$A = \{aX\}, \qquad B = \{bX\}, \qquad C = \{cX\}$$

so that the points $(a)$, $(b)$, $(c)$ have equations $A = 0$, $B = 0$, $C = 0$, we find that the point $(\lambda a + \mu b + \nu c)$ has the equation

$$\lambda A + \mu B + \nu C = 0$$

Thus the barycentric coordinates of any point are simply the coefficients of $A$, $B$, $C$ when its equation is expressed in terms of the equations of those points. The above remarks serve to justify Möbius's "barycentric calculus" (so effectively used by Baker), where the general point is denoted by

$$\lambda A + \mu B + \nu C$$

(with "$= 0$" omitted). In this notation our triangle of reference (formed by the points $X_1 = 0$, $X_2 = 0$, $X_3 = 0$) is simply $X_1 X_2 X_3$, and the point $(x)$ is $x_1 X_1 + x_2 X_2 + x_3 X_3$. Thus

$$(x_1, x_2, x_3) = x_1(1, 0, 0) + x_2(0, 1, 0) + x_3(0, 0, 1)$$

as in vector analysis. (In fact, if we think of the plane as lying in an affine space, we may interpret these symbols as vectors leading from some fixed origin outside the plane to the points considered; then the expression $\lambda A + \mu B + \nu C$ is a sum of vectors.) Historically, this barycentric calculus (1827) preceded Plücker's line coordinates (1828–1830). Von Staudt's synthetic approach to projective geometry came later still, as we have seen. Grassmann, a contemporary of von Staudt, developed a "calculus of extension" in which both points and lines are represented as vectors: the vector product of two points is their join, and the vector product of two lines is their intersection (cf. Sec. 12·3).

One very practical rule emerges from this little digression. When seeking an analytic proof for a theorem concerning a triangle, we are justified in taking this as triangle of reference; and any fixed point not on a side of the triangle may be named $(1, 1, 1)$. Thus Exercise 2 of Sec. 12·3 would suffice for a proof of Desargues' theorem (our 2·25) and Exercise 3 for a proof of Pappus's theorem (our 4·31). (This is far neater than the proofs of Pappus's theorem given in

most textbooks on analytic geometry.) For theorems involving a
quadrangle it is often convenient to take the vertices to be $(1, \pm 1, \pm 1)$,
so that the six sides are $x_i \pm x_j = 0$ $(i < j)$ and the diagonal triangle
is the triangle of reference. Dually, a given quadrilateral may be taken
to have sides $[1, \pm 1, \pm 1]$ and vertices $X_i \pm X_j = 0$.

The following special collineations will be found useful: a homology with center $(0, 0, 1)$ and axis $[0, 0, 1]$,

**12·54** $$x_1' = x_1, \qquad x_2' = x_2, \qquad x_3' = \frac{x_3}{\rho}$$

and the elation with center $(c_1, c_2, 0)$ and axis $[0, 0, 1]$,

**12·55** $$x_1' = x_1 + c_1 x_3, \qquad x_2' = x_2 + c_2 x_3, \qquad x_3' = x_3$$

### EXERCISES

1. Give an analytic proof for Sec. 2·4, Exercise 1.

2. Find the collineations that transform $(1, 0, 0)(0, 1, 0)(0, 0, 1)(1, 1, 1)$
into the following quadrangles:

(i)            $(1, 0, 0)(0, 1, 0)(0, 0, 1)(a_1, a_2, a_3)$

(ii)          $(-1, 1, 1)(1, -1, 1)(1, 1, -1)(1, 1, 1)$

(iii)         $(0, 1, 0)(0, 0, 1)(1, 0, 0)(1, 1, 1)$

(iv)         $(0, 1, 0)(0, 0, 1)(1, 1, 1)(1, 0, 0)$

The last two collineations are periodic. What are their periods?

3. Find the collineation that interchanges $(\pm 1, 1, 1)$ and also interchanges $(\pm 1, -1, 1)$ (cf. 5·31). Where are the center and axis of this
harmonic homology?

4. Give an analytic proof for the exercise to Sec. 5·3.

5. Find the elation with axis $[1, 0, 0]$ transforming $(1, 0, 0)$ into $(1, 1, 0)$
(cf. 5·22). Where is its center?

6. Find the homology or elation that transforms the triangle of reference
into $(k_1, 1, 1)(1, k_2, 1)(1, 1, k_3)$ (cf. 5·24). When will it be an elation?

7. Express the following three collineations in terms of line coordinates.
Find their invariant points and lines.

(i)    $x_1' = x_1,$       $x_2' = x_3,$        $x_3' = -x_2$

(ii)   $x_1' = c_{11} x_1,$    $x_2' = x_2,$       $x_3' = c_{32} x_2 + x_3$

(iii)   $x_1' = x_1,$       $x_2' = c_{21} x_1 + x_2$    $x_3' = c_{31} x_1 + c_{32} x_2 + x_3$

**12·6   The General Polarity.**   Since the product of any two correlations is a collineation, the general correlation can be obtained by combining the general collineation 12·51 with the special correlation that
interchanges $X_i'$ and $x_i'$, thus:

$$\begin{cases} X'_i = \Sigma c_{ij}x_j & (i = 1, 2, 3) \\ X_j = \Sigma c_{ij}x'_i & (j = 1, 2, 3) \end{cases}$$

(All incidences are dualized, as $\Sigma x'_i X'_i = \Sigma X_j x_j$.)

This correlation is a *polarity* if it is equivalent to the inverse correlation $X'_j = \Sigma c_{ij}x_i$ or (interchanging $i$ and $j$)

$$X'_i = \Sigma c_{ji}x_j \qquad (i = 1, 2, 3)$$

This means that $c_{ji} = \lambda c_{ij}$, with the same $\lambda$ for all $i$ and $j$, so that $c_{ij} = \lambda c_{ji} = \lambda^2 c_{ij}$, $\lambda^2 = 1$, $\lambda = \pm 1$. But we cannot have $\lambda = -1$, as that would make

$$\Delta = \begin{vmatrix} 0 & c_{12} & -c_{31} \\ -c_{12} & 0 & c_{23} \\ c_{31} & -c_{23} & 0 \end{vmatrix} = 0$$

Hence $\lambda = 1$, and $c_{ji} = c_{ij}$. In other words, a correlation is a polarity if and only if the matrix of coefficients $c_{ij}$ is symmetric.

To emphasize this extra condition we shall write $a_{ij}$ ($=a_{ji}$) instead of $c_{ij}$. Moreover, the nature of a polarity is such that no confusion can be caused by omitting the prime ['] and writing simply

**12·61** $$X_i = \Sigma a_{ij}x_j \qquad (i = 1, 2, 3)$$

These equations give us the polar $[X]$ of a given point $(x)$. Solving them, we obtain the pole $(x)$ of a given line $[X]$ in the form

**12·62** $$x_i = \Sigma A_{ij}X_j \qquad (i = 1, 2, 3)$$

and we know that the coefficients are connected as follows:

$$a_{ji} = a_{ij}, \qquad A_{ji} = A_{ij}, \qquad \Sigma a_{ij}A_{ik} = \delta_{jk}$$

so that

$$\det (a_{ij}) = \Delta \neq 0 \qquad \text{and} \qquad \det (A_{ij}) = \Delta^{-1}$$

Two points $(x)$ and $(y)$ are *conjugate* (Sec. 5·5) if $(x)$ lies on the polar $[Y]$ of $(y)$. Since $Y_i = \Sigma a_{ij}y_j$, the condition $\{xY\} = 0$ or $\Sigma x_i Y_i = 0$ becomes

$$\Sigma\Sigma a_{ij}x_iy_j = 0$$

Letting $(x)$ vary, we see that this is *the equation for the polar of* $(y)$. We shall often write it in the abbreviated form

**12·63** $$(xy) = 0$$

Dually, the condition for lines $[X]$ and $[Y]$ to be conjugate, or the equation for the pole of $[Y]$, is

**12·64**                    $$[XY] = 0$$

where $[XY] = \Sigma\Sigma A_{ij}X_iY_j$.

The fact that a polarity induces an *involution* of conjugate points on any non-self-conjugate line (5·53) may be verified by writing down the condition for points $(x, 1, 0)$ and $(x', 1, 0)$ (on the fixed line $[0, 0, 1]$) to be conjugate, *viz.*,

$$a_{11}xx' + a_{12}(x + x') + a_{22} = 0$$

As we saw at the top of page 152, this is a proper involution unless

$$a_{11}a_{22} - a_{12}^2 = 0$$

in which case $A_{33} = 0$, and the line $[0, 0, 1]$ is self-conjugate.

For an analytic proof of Chasles's theorem (our 5·61), we apply the general polarity 12·61 to the vertices of the triangle of reference, obtaining the sides

$$[a_{11}, a_{21}, a_{31}], \qquad [a_{12}, a_{22}, a_{32}], \qquad [a_{13}, a_{23}, a_{33}]$$

of another triangle. Since the result is trivial when a pair of corresponding sides coincide, we may assume that at least two of $a_{23}$, $a_{31}$, $a_{12}$ are different from zero. Then any two corresponding sides are concurrent with

$$[a_{31}a_{12}, a_{12}a_{23}, a_{23}a_{31}]$$

For von Staudt's converse theorem (our 5·71), we observe that the sides of the triangle of reference are related to the points

$$(k_1, 1, 1), \qquad (1, k_2, 1), \qquad (1, 1, k_3)$$

by the polarity 12·62 with $A_{ii} = k_i$ and every other $A_{ij} = 1$, namely,

$$x_1 = k_1X_1 + X_2 + X_3$$
$$x_2 = X_1 + k_2X_2 + X_3$$
$$x_3 = X_1 + X_2 + k_3X_3$$

(see Sec. 12·3, Exercise 2).

Returning to the polarity 12·61 or 12·63, we observe that the condition for $(0, 1, 0)$ and $(0, 0, 1)$ to be conjugate is $a_{23} = 0$. Thus the triangle of reference is self-polar if and only if

$$a_{23} = a_{31} = a_{12} = 0$$

By choosing such a triangle of reference we reduce a polarity to its
*canonical form*

$$X_i = a_{ii}x_i \quad (i = 1, 2, 3;\ a_{11}a_{22}a_{33} = \Delta \neq 0)$$

The coefficients $a_{ii}$ are determined by one further pole and polar, as in
5·63. In fact, if $ABC$ is the triangle of reference while $P$ is $(C_1, C_2, C_3)$
and $p$ is $[c_1, c_2, c_3]$, then the polarity $(ABC)$ $(Pp)$ is of the above form
with $a_{ii} = c_i/C_i$. In particular, the canonical polarity relating $(1, 1, 1)$
to $[c_1, c_2, c_3]$ is

**12·65**　　　　　　　　　　$$X_i = c_i x_i \quad (i = 1, 2, 3;\ c_1 c_2 c_3 \neq 0)$$

So far, we have insisted that the determinant $\Delta$ shall not vanish.
It is interesting to see what kind of degenerate polarity remains if we
allow $\Delta = 0$. The relations 12·61 still provide a unique line $[X]$ for each
point $(x)$. But now all polars $[X]$ pass through one fixed point, and each
is the polar of infinitely many $(x)$'s. In fact, the vanishing of the
determinant implies the existence of numbers $z_1$, $z_2$, $z_3$, not all zero,
such that

$$\Sigma a_{ij}z_i = 0 \qquad (j = 1, 2, 3)$$

Hence, for any point $(x)$, $\Sigma\Sigma a_{ij}z_i x_j = 0$; which means that the polar
$[X]$, satisfying $\Sigma z_i X_i = 0$, always passes through a certain point $(z)$.
Such a line $[X]$ is also the polar of $(x + \nu z)$ for any $\nu$.

In other words, when $\Delta = 0$, there is a point $(z)$ that is conjugate
to every point $(x)$. This universal conjugate point $(z)$ is unique unless
all points have the same polar $[X]$. This completely degenerate case
arises when $a_{ij}$ is of the form $a_i a_j$, so that 12·61 reduces to

$$X_i = a_i \Sigma a_j x_j$$

This means that all cofactors of order 2 in $\Delta$ vanish, or that the matrix
$(a_{ij})$ is of rank 1.

If $(a_{ij})$ is of rank 2, so that $(z)$ is unique, the condition for two
points $(x)$ and $(y)$ to be conjugate is still 12·63 and there is still an
involution of conjugate points on any line not passing through $(z)$.
Of such a pair of points, each is joined to $(z)$ by the polar of the other;
thus we have an involution of conjugate lines through $(z)$. For each
pair of lines in this involution the polarity relates every point on either
line to the other line. In this sense, the degenerate polarity *is* the
involution of conjugate lines through $(z)$.

Dually, the transformation 12·62 with $\det(A_{ij}) = 0$ represents
another kind of degenerate polarity, such that the pole of any line lies

on one fixed line $[Z]$. If the matrix $(A_{ij})$ is of rank 1, we have complete degeneracy: all lines have the same pole. But if it is of rank 2, the line $[Z]$ is unique and the degenerate polarity is essentially an involution of point pairs on $[Z]$.

The above remarks may be summarized as follows:

**12·66**  *If $(a_j)$ is a matrix of rank 2, the polarity $X_i = \Sigma a_{ij}x_j$ degenerates into an involution of line pairs through a fixed point. Dually, if $(A_{ij})$ is of rank 2, the polarity $x_i = \Sigma A_{ij}X_j$ degenerates into an involution of point pairs on a fixed line.*

### EXERCISES

1. Prove Hesse's theorem (our 5·54), using the general polarity and the quadrilateral $[1, \pm 1, \pm 1]$, whose vertices are

$$(0, 1, \pm 1), \qquad (\pm 1, 0, 1), \qquad (1, \pm 1, 0)$$

2. Prove 5·62, using the triangle of reference.

3. Verify 6·21 and 6·22 as applied to the triangle of reference and the unit line $[1, 1, 1]$.

4. Solve Exercise 6 of Sec. 7·7, using the triangle of reference.

5. Use the following coordinates in Sec. 5·7, Exercise 2:

$$A\ (1, 0, 0), \qquad B\ (0, 1, 0), \qquad C'\ (0, 0, 1)$$
$$L\ (0, 1, p), \qquad M\ (q, 0, 1), \qquad C\ (1, r, 0)$$
$$B'N\ [0, 1, P], \qquad A'N\ [Q, 0, 1], \qquad A'B'\ [1, R, 0]$$

The incidences in Fig. 4·3A require

$$1 + qQ + (rR)^{-1} = 0, \qquad 1 + rR + (pP)^{-1} = 0$$

Verify that these equations imply $1 + pP + (qQ)^{-1} = 0$ (thus providing an alternative proof for Pappus's theorem) and $pqrPQR = 1$. Finally, obtain the condition $pqr = PQR = 1$ for the reciprocity suggested by the above notation to be induced by a polarity.

6. Verify that the relations

$$X_1 = \lambda c_1' x_1, \qquad X_2 = c_2 x_2, \qquad X_3 = (c_3 + \lambda c_3')x_3$$

define a pencil of polarities transforming the unit point $(1, 1, 1)$ into the pencil of lines concurrent with $[0, c_2, c_3]$ and $[c_1', 0, c_3']$ and that this is a self-dual system (*i.e.*, a range as well as a pencil) if $c_3' = 0$. Setting $c_1' = 1$, we thus obtain the system

$$X_1 = \lambda x_1, \qquad X_2 = c_2 x_2, \qquad X_3 = c_3 x_3$$

(cf. 5·82). Verify that the locus of poles of $[1, 1, 1]$ is the line $c_2 x_2 = c_3 x_3$.

7. Find the locus of poles of a fixed line $[X]$ wo the polarities

$$X_1 = x_3 + \lambda x_1, \qquad X_2 = x_2, \qquad X_3 = x_1$$

(*Hint:* Rewrite these relations as $\rho X_1 = x_3 + \lambda x_1$, $\rho X_2 = x_2$, $\rho X_3 = x_1$; then eliminate $\rho$ and $\lambda$.)

**12·7  Conics.** The condition for a point $(x)$ to the self-conjugate for the polarity 12·61 is $(xx) = 0$, or

$$a_{11}x_1^2 + a_{22}x_2^2 + a_{33}x_3^2 + 2a_{23}x_2x_3 + 2a_{31}x_3x_1 + 2a_{12}x_1x_2 = 0$$

Hence the polarity is elliptic or hyperbolic according as the quadratic form $(xx)$ (with determinant $\Delta \neq 0$) is definite or indefinite.* In the latter case the locus of self-conjugate points is the conic†

$$(xx) = 0$$

and the envelope of self-conjugate lines is the same conic in the form

$$[XX] = 0$$

In particular, the condition for $(x)$ to be self-conjugate for 12·65 is $\Sigma c_i x_i^2 = 0$; thus the canonical polarity is elliptic or hyperbolic according as the three nonvanishing coefficients $c_i$ do or do not have the same sign, and in the latter case the conic is

$$c_1x_1^2 + c_2x_2^2 + c_3x_3^2 = 0 \qquad \text{or} \qquad \frac{X_1^2}{c_1} + \frac{X_2^2}{c_2} + \frac{X_3^2}{c_3} = 0$$

By the coordinate transformation

$$x_i \to |c_i|^{-\frac{1}{2}}x_i \qquad X_i \to |c_i|^{\frac{1}{2}}X_i$$

we can reduce the coefficients to $\pm 1$. Then, renumbering the three coordinates if necessary, 12·65 becomes

**12·71**            $$X_1 = x_1, \qquad X_2 = x_2, \qquad X_3 = \pm x_3$$

with the upper or lower sign according as the polarity is elliptic or hyperbolic. [In the former case this amounts to taking $(1, 1, 1)$ to be one of the four points described in Sec. 7·7, Exercise 6.] Thus any conic may be expressed as

**12·72**         $$x_1^2 + x_2^2 - x_3^2 = 0, \qquad X_1^2 + X_2^2 - X_3^2 = 0$$

In this form, the triangle of reference is self-polar. Another useful equation,

**12·73**                          $$x_2^2 - x_3x_1 = 0$$

---

* A necessary and sufficient condition for a *definite* form (or *elliptic* polarity) is that the three numbers $a_{11}$, $A_{22}$, $\Delta$ all have the same sign. See Veblen and Young (Ref. 43, p. 205).

† Hesse (Ref. 17) seems to have been the first to write the equation for a conic in the form $\Sigma\Sigma a_{ij}x_ix_j = 0$.

(where the triangle of reference is formed by two tangents and the join of their points of contact), is derived from 12·72 by the transformation

$$x_1 \rightarrow \tfrac{1}{2}(x_1 - x_3), \qquad x_2 \rightarrow x_2, \qquad x_3 \rightarrow \tfrac{1}{2}(x_1 + x_3)$$

This exhibits the conic as the locus of the point of intersection of the projectively related lines $x_1 - tx_2 = 0$ and $x_2 - tx_3 = 0$, as in 6·54. In other words, the conic is the locus of the point $(t^2,\ t,\ 1)$ whose *parameter* is $t$.

Since the condition for the conic $(xx) = 0$ to pass through $(1, 0, 0)$ is $a_{11} = 0$, the general conic circumscribing the triangle of reference is

$$a_{23}x_2x_3 + a_{31}x_3x_1 + a_{12}x_1x_2 = 0 \qquad (a_{23}a_{31}a_{12} \neq 0)$$

The coordinate transformation

$$x_1 \rightarrow a_{23}x_1, \qquad x_2 \rightarrow a_{31}x_2, \qquad x_3 \rightarrow a_{12}x_3$$

converts this into

**12·74**
$$x_2x_3 + x_3x_1 + x_1x_2 = 0$$

or $x_1^{-1} + x_2^{-1} + x_3^{-1} = 0$. This exhibits the conic as the locus of trilinear poles of lines through the unit point $X_1 + X_2 + X_3 = 0$ (which is the point described in Sec. 7·7, Exercise 5). Working out the cofactors in the determinant, we obtain the envelope equation

$$X_1^2 + X_2^2 + X_3^2 - 2X_2X_3 - 2X_3X_1 - 2X_1X_2 = 0$$

or $X_1^{\frac{1}{2}} \pm X_2^{\frac{1}{2}} \pm X_3^{\frac{1}{2}} = 0$. Dually, a conic inscribed in the triangle of reference is

**12·75**    $X_2X_3 + X_3X_1 + X_1X_2 = 0$    or    $x_1^{\frac{1}{2}} \pm x_2^{\frac{1}{2}} \pm x_3^{\frac{1}{2}} = 0$

If $\Delta = 0$, the equation $(xx) = 0$ represents a *degenerate conic*, the locus of self-conjugate points for a degenerate polarity (see page 173. One such point is $(z)$. If $(y)$ is another, every point collinear with $(y)$ and $(z)$ must be likewise self-conjugate, since

$$\Sigma \Sigma a_{ij}(y_i + vz_i)(y_j + vz_j) = (yy) + 2v(yz) + v^2(zz) = 0$$

In the completely degenerate case when $a_{ij} = a_i a_j$, the equation $(xx) = 0$ reduces to $\{ax\}^2 = 0$, which is essentially the line $[a]$. Thus the equation $(xx) = 0$ with $\det(a_{ij}) = 0$ represents a degenerate conic locus consisting of a single point [for example, $x_1^2 + x_2^2 = 0$, which is $(0, 0, 1)$] or a line (for example, $x_1^2 = 0$) or two lines (for example, $x_1x_2 = 0$); but not more than two, or the equation would involve the product of three or more linear factors.

Dually, the equation $[XX] = 0$ with $\det(A_{ij}) = 0$ represents a degenerate conic envelope consisting of a line (for example, $X_1^2 + X_2^2 = 0$ which is $[0, 0, 1]$) or a point (for example, $X_1^2 = 0$) or two points (for example, $X_1 X_2 = 0$).*

The following result is typical of many applications of the $(xy)$ notation. The condition for the join of two points $(x)$ and $(y)$ to be a tangent to the conic $(xx) = 0$ is

**12·76**     $$(xx)(yy) - (xy)^2 = 0$$

To see this, let $(x + \mu y)$ be the point of contact of such a tangent. The point of contact must be conjugate to both $(x)$ and $(y)$; hence

$$(x\ x + \mu y) = 0, \qquad (x + \mu y\ y) = 0$$

i.e.,

$$(xx) + \mu(xy) = 0, \qquad (xy) + \mu(yy) = 0$$

We obtain 12·76 by eliminating $\mu$; and the argument can be reversed. The same equation may be obtained as the degeneracy condition for the involution

$$(xx) + (xy)(\mu + \mu') + (yy)\mu\mu' = 0$$

of conjugate points $(x + \mu y)$ and $(x + \mu' y)$ on the line $(x)(y)$.

If $(y)$ is a fixed exterior point, 12·76 is the combined equation for the two tangents that can be drawn to the conic from that point. Dually, the conic $[XX] = 0$ meets a secant $[Y]$ in the two points

$$[XX][YY] - [XY]^2 = 0$$

For problems involving two conics, it is convenient to use the notation $(xx)' = \Sigma\Sigma a'_{ij}x_ix_j$. If two conics $(xx) = 0$ and $(xx)' = 0$ have four points of intersection, the equation

**12·77**     $$(xx) + \lambda(xx)' = 0$$

represents a conic (possibly degenerate) through the same four points. Moreover, this is the most general conic of the quadrangular pencil; for if $(y)$ is any fifth point on such a conic, we merely have to choose $\lambda$ so as to satisfy $(yy) + \lambda(yy)' = 0$. Taking a more general standpoint, let us call 12·77 a *pencil of conics* even if the conics have no common self-polar triangle (so that the definition implied in 6·82 cannot be used). We shall find that most of the theorems about pencils remain

---

* The tangents to a very small circle visibly resemble a pencil of lines, enveloping a single point; and the tangents to a very flat ellipse (with eccentricity nearly 1) resemble two such pencils. See Robson (Ref. 33, p. 67).

valid. Extending 5·81, we observe that the polars of $(y)$ form the pencil of lines

$$(xy) + \lambda(xy)' = 0$$

whose center, given by solving the two equations $(xy) = (xy)' = 0$ for $(x)$, is conjugate to $(y)$ wo every one of the conics. As for 6·81: the poles of a line $(y)(z)$ all satisfy the equation

**12·78**              $(xy)(xz)' - (xz)(xy)' = 0$

which is obtained by eliminating $\lambda$ from

$$(xy) + \lambda(xy)' = 0, \qquad (xz) + \lambda(xz)' = 0$$

The common conjugates of points $(\mu y + z)$ on the line $(y)(z)$ satisfy the same equation, obtained by eliminating $\mu$ from

$$\mu(xy) + (xz) = 0, \qquad \mu(xy)' + (xz)' = 0$$

Desargues' involution theorem (our 6·73 and 6·59) can be extended as follows: The conic 12·77 passes through $(\mu y + z)$ if

$$\mu^2(yy) + 2\mu(yz) + (zz) + \lambda[\mu^2(yy)' + 2\mu(yz)' + (zz)'] = 0$$

By considering the sum and products of the roots of this quadratic equation for $\mu$, we see that two points $(\mu_1 y + z)$ and $(\mu_2 y + z)$ lie on the same conic of the pencil if

$$(\mu_1 + \mu_2)[(yy) + \lambda(yy)'] = -2[(yz) + \lambda(yz)']$$

and

$$\mu_1\mu_2[(yy) + \lambda(yy)'] = (zz) + \lambda(zz)'$$

Eliminating $\lambda$, we obtain the equation

$$2\mu_1\mu_2[(yy)(yz)' - (yz)(yy)'] + (\mu_1 + \mu_2)[(yy)(zz)' - (zz)(yy)'] \\ + 2[(yz)(zz)' - (zz)(yz)'] = 0$$

which is also the condition for the points $(\mu_1 y + z)$ and $(\mu_2 y + z)$ to be conjugate wo 12·78. Hence:

*Those conics of the pencil 12·77 that meet the line $(y)(z)$ do so in pairs of points that are conjugate wo the conic 12·78.*

In the case of a quadrangular pencil, there are three values of $\lambda$ for which 12·77 consists of a line pair. This fact is neatly employed in the following proof* of Pascal's theorem (our 7·21). Using the notation of Fig. 7·2A, let the lines $BB'$, $CA'$, and $AC'$ be $[X]$, $[Y]$, and $[Z]$. Then

* Robson (Ref. 34, p. 91).

for a certain $\lambda$ the equation

$$(xx) + \lambda\{Xx\}\{Yx\} = 0$$

represents the line pair $BA'$, $CB'$; and for a certain $\mu$ the equation

$$(xx) + \mu\{Xx\}\{Zx\} = 0$$

represents the pair $BC'$, $AB'$. Subtracting these, we obtain another degenerate conic

$$\{Xx\}\,(\lambda\{Yx\} - \mu\{Zx\}) = 0$$

through the common points $B$, $B'$, $N$, $L$ of the first two. Now, the first line pair is $BN$, $LB'$, and the second is $BL$, $NB'$; hence the third must be $BB'$, $NL$. The factor $\{Xx\}$ gives the line $[X]$ which is $BB'$; therefore $NL$ is $[\lambda Y - \mu Z]$, concurrent with $CA'$ and $AC'$.

An interesting special case of 12·78 arises when the four common points are $(1, \pm1, \pm1)$, so that the quadrangular pencil of conics is given by

$$c_1x_1^2 + c_2x_2^2 + c_3x_3^2 = 0, \qquad c_1 + c_2 + c_3 = 0$$

for various values of the $c$'s. The polar of a fixed point $(x)$ is

$$[c_1x_1,\ c_2x_2,\ c_3x_3]$$

which continually passes through the fixed point $(x_1^{-1}, x_2^{-1}, x_3^{-1})$. This "quadratic transformation" $(x) \to (x^{-1})$ (which somewhat resembles inversion wo a circle) transforms the points of a line $[X]$ into the points of a conic

$$X_1x_1^{-1} + X_2x_2^{-1} + X_3x_3^{-1} = 0$$

through the vertices of the triangle of reference. This conic is also the locus of poles of $[X]$, as we see by eliminating the $c$'s from

$$X_i = c_ix_i \qquad \text{and} \qquad \Sigma c_i = 0$$

## EXERCISES

1. Verify 6·43 as applied to the quadrangle $(1, \pm1, \pm1)$.

2. Show that a unique conic can be drawn through $(1, 1, 1)$ to touch $[0, 0, 1]$ at $(1, 0, 0)$ and $[1, 0, 0]$ at $(0, 0, 1)$ (cf. 6·53 and 12·73).

3. Verify Sec. 6·5, Exercise 3, taking the exterior points $p \cdot s$ and $q \cdot r$ to be $(y)$ and $(z)$. The conic through the six points turns out to be

$$(xx)(yz) - (xy)(xz) = 0$$

**4.** If $[X]$ meets $(xx) = 0$ in two points, prove that the two lines joining these points to another point $(y)$ are given by

$$\{Xy\}^2(xx) - 2\{Xx\}\{Xy\}(xy) + \{Xx\}^2(yy) = 0$$

*Hint:* What value of $\mu$ will make $(x + \mu y)$ lie on the conic?

**5.** If the sides of a variable triangle pass through three fixed points $(\lambda, 1, 0)$, $(1, \mu, 0)$, $(1, 1, \nu)$, while the vertices opposite to the first two sides run along the respective lines $[1, 0, 0]$, $[0, 1, 0]$, prove that the third vertex will trace a conic or a line (as in 6·61 or the Exercise to Sec. 4·2). *Hint:* Take the triangle to be

$$(0, x_2 - \mu x_1, x_3)(x_1 - \lambda x_2, 0, x_3)(x_1, x_2, x_3)$$

**6.** Show that the conics

$$(xx) + \lambda(xy)^2 = 0$$

form a self-dual system (Sec. 6·8; cf. 12·76). What happens when $(y)$ is an interior point?

**7.** Considering the conic 12·73 as the locus of $(t^2, t, 1)$, prove that the secant joining the points with parameters $t$ and $t'$ is $[1, -(t + t'), tt']$ and that the tangent at the point $t$ is $[1, -2t, t^2]$. Deduce the envelope equation

$$X_2^2 - 4X_3X_1 = 0$$

and check this by direct computation of cofactors.

**8.** Show that the quadratic transformation $(x) \rightarrow (x^{-1})$ transforms a conic through two vertices of the triangle of reference into a conic through the same two vertics. How many conics are transformed into themselves? (Six pencils.)

**9.** Give an analytic treatment of Sec. 7·2, Exercise 6, taking the conic in the form 12·74, with $ABC$ for triangle of reference. *Hint:* Take $A_1B_1C_1$ to be $(\lambda, 1, -1)(-1, \mu, 1)(1, -1, \nu)$. The point of concurrence turns out to be $(\lambda, \mu, \nu)$.

**12·8   The Affine Plane: Affine and Areal Coordinates.** We saw, in Sec. 8·1, how the affine plane can be derived from the projective plane by removing one line. Analytically, the simplest way to do this is to remove one side of the triangle of reference, say $[0, 0, 1]$ or $x_3 = 0$. The remaining two sides are then called coordinate *axes*. Any point for which $x_3 \neq 0$ can be normalized (dividing through by $x_3$) so as to take the form $(x_1, x_2, 1)$, which can then be abbreviated to $(x_1, x_2)$. In this manner we obtain a unique symbol for every ordinary point. These nonhomogeneous coordinates $x_1$, $x_2$ are called *affine* coordinates.

Lines $[X_1, X_2, X_3]$, where $X_1$ and $X_2$ are fixed while $X_3$ varies, are parallel, since they are concurrent with $[X_1, X_2, 0]$ and $[0, 0, 1]$. In particular, the line $[1, 0, -x_1]$ is parallel to the axis $[1, 0, 0]$ and

$[0, 1, -x_2]$ to $[0, 1, 0]$. These four lines form a parallelogram whose vertices are

$$(0, 0), \quad (x_1, 0), \quad (0, x_2), \quad (x_1, x_2)$$

as in Fig. 12·8A.

From the remark at the end of Sec. 11·7, $x_1$ is the *abscissa* of $(x_1, 0, 1)$, referred to

$$P_0 = (0, 0, 1), \quad P_1 = (1, 0, 1), \quad P_\infty = (1, 0, 0)$$

Comparing this with Sec. 8·4, we see that $x_1$ is actually the *distance* from $(0, 0)$ to $(x_1, 0)$, in terms of the distance to $(1, 0)$ as unit. Similarly

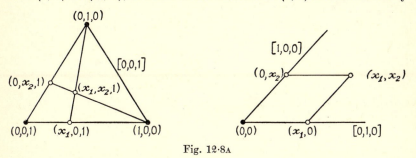

Fig. 12·8A

$x_2$ is the distance from $(0, 0)$ to $(0, x_2)$ in terms of the distance to $(0, 1)$ as unit. In affine geometry these two units are, of course, independent; it would be meaningless to regard them as being "equal."

The affine theory of conics can be developed by choosing the coordinate axes in convenient positions: *e.g.*, an ellipse for which the coordinate axes are conjugate diameters can be taken in the form

**12·81** $$x_1^2 + x_2^2 = 1$$

while a parabola touching the axis $x_1 = 0$ at $(0, 0)$ is

$$x_2^2 = 2x_1$$

and a hyperbola whose asymptotes are the coordinate axes is

$$x_1 x_2 = 1$$

The first of these three equations reminds us of a circle. This is natural enough when we remember that all the *affine* properties of the circle are properties of the ellipse. We may even regard the ellipse as the locus of $(\cos t, \sin t)$, interpreting $t$ as twice the area of the sector from $(1, 0)$ to $(\cos t, \sin t)$. Similarly, the hyperbola

$$x_1^2 - x_2^2 = 1$$

(for which the coordinate axes are a pair of conjugate diameters) is the locus of (cosh $t$, sinh $t$), where $t$ is twice the area of the sector from (1, 0) to that point (as we may easily verify by integration).

Instead of [0, 0, 1] we might take the line at infinity to be [1, 1, 1]. Then (1, 1, 1), the trilinear pole of this line, is the centroid of the triangle of reference, and (0, 1, 1), etc., are the mid-points of the sides. It follows that the ratio of the distances of (0, $x_2$, $x_3$) from (0, 1, 0) and (0, 0, 1) (measured in opposite directions) is $x_3/x_2$, and that the areas of the three triangles joining ($x_1$, $x_2$, $x_3$) to the sides of the triangle of reference are proportional to $x_1:x_2:x_3$. Accordingly, these are called *areal* coordinates. [By a well-known result in vector analysis, ($x$) is the centroid of masses $x_1$, $x_2$, $x_3$ at the vertices of the triangle of reference.]

Since the line at infinity is $x_1 + x_2 + x_3 = 0$, any ordinary point can be normalized so that $x_1 + x_2 + x_3 = 1$. The effect is the same as if we took affine coordinates in three dimensions (analogous to the two-dimensional affine coordinates defined above), restricting attention to the plane whose equation is

$$x_1 + x_2 + x_3 = 1$$

Then the vectors $X_1$, $X_2$, $X_3$ (page 169) proceed along the three coordinate axes.

We have seen how the areal coordinates of a point may be measured as areas. But what about the areal coordinates of a line?

**12·82** *The areal coordinates of a line are proportional to its distances from the three vertices of the triangle of reference.*[*]

*Proof:* By similar triangles, these distances have the same ratios in whatever directions they are measured, provided that the direction is the same for all three vertices. In particular, we may measure them along the side [1, 0, 0]. Since [$X_1$, $X_2$, $X_3$] meets [1, 0, 0] in the point (0, $X_3$, $-X_2$), its distances from (0, 1, 0) and (0, 0, 1) are $-X_2:X_3$ measured in opposite directions, or $X_2:X_3$ measured in the same direction. By symmetry, the distances from all three vertices, measured in the same direction, are $X_1:X_2:X_3$.

This agrees nicely with the fact that we obtain a parallel line if we increase $X_1$, $X_2$, $X_3$ all by the same amount.

### EXERCISES

1. Show that the affine equation

$$a_{11}x_1^2 + 2a_{12}x_1x_2 + a_{22}x_2^2 = 1 \quad (\Delta = a_{12}^2 - a_{11}a_{22})$$

[*] Salmon (Ref. 38, p. 11).

represents a hyperbola if $\Delta > 0$, two parallel lines if $\Delta = 0$; but if $\Delta < 0$ it represents an ellipse or nothing according as $a_{11}$ is positive or negative. In the case of a hyperbola, what is the combined equation for the asymptotes?

2. Find the locus and envelope areal equations of the parabola that touches $[0, 0, 1]$ at $(1, 0, 0)$ and $[1, 0, 0]$ at $(0, 0, 1)$.

3. Show that the areal equation

$$c_1 x_1^2 + c_2 x_2^2 + c_3 x_3^2 = 0$$

represents a parabola if

$$c_1^{-1} + c_2^{-1} + c_3^{-1} = 0$$

4. Show that $(A_{11} + A_{12} + A_{13}, \; A_{21} + A_{22} + A_{23}, \; A_{31} + A_{32} + A_{33})$ is the center of a nondegenerate conic whose areal equation is $(xx) = 0$ or $[XX] = 0$. Show further that the conic is a hyperbola if $\Delta$ has the same sign as $A_{11} + A_{22} + A_{33} + 2A_{23} + 2A_{31} + 2A_{12}$, in which case the asymptotes are given by

$$(A_{11} + A_{22} + A_{33} + 2A_{23} + 2A_{31} + 2A_{12})(xx) = (x_1 + x_2 + x_3)^2$$

### 12·9 The Euclidean Plane: Cartesian and Trilinear Coordinates.

We saw, in Sec. 9·1, how a Euclidean metric can be introduced into the affine plane by singling out an elliptic involution in the line at infinity, with the result that a circle has this for its involution of conjugate points on that line. In particular, we derive *rectangular Cartesian* coordinates from affine coordinates by calling 12·81 a circle of radius 1, so that the absolute involution relates $(x_1, x_2, 0)$ and $(y_1, y_2, 0)$, where

**12·91** $$x_1 y_1 + x_2 y_2 = 0$$

The dilatation 12·54 transforms 12·81 into a circle of radius $\rho$:

$$x_1^2 + x_2^2 = \rho^2$$

The translation 12·55 (with $x_3 = 1$) transforms this into the general circle

$$(x_1 - c_1)^2 + (x_2 - c_2)^2 = \rho^2$$

as in classical analytic geometry.

Areal coordinates belong, as we have seen, to the affine plane. They may still be employed after the introduction of a Euclidean metric (when the triangle of reference acquires definite lengths for its sides, say $a$, $b$, $c$); but for many purposes it is desirable to make the transformation

$$x_1 \rightarrow a x_1, \qquad x_2 \rightarrow b x_2, \qquad x_3 \rightarrow c x_3$$

to *trilinear* coordinates, which are proportional to the distances of the point considered from the three sides, so that the unit point is the incenter (instead of the centroid) and the line at infinity is [a, b, c].

In terms of the angles $A$, $B$, $C$ of the triangle of reference $ABC$, any point on the perpendicular from $A$ to $BC$ is at distances from $CA$ and $AB$ which are proportional to $\cos C : \cos B$. Since

$$a - b \cos C - c \cos B = 0$$

the point at infinity on this altitude line is $(1, -\cos C, -\cos B)$, and the absolute involution is the degenerate polarity

**12·92**
$$\begin{cases} x_1 = X_1 - X_2 \cos C - X_3 \cos B \\ x_2 = -X_1 \cos C + X_2 - X_3 \cos A \\ x_3 = -X_1 \cos B - X_2 \cos A + X_3 \end{cases}$$

which relates $[1, 0, 0]$ to $(1, -\cos C, -\cos B)$, and so on. Thus the condition for two lines $[X]$ and $[Y]$ to be perpendicular is*

$$X_1 Y_1 + X_2 Y_2 + X_3 Y_3 - (X_2 Y_3 + X_3 Y_2) \cos A$$
$$- (X_3 Y_1 + X_1 Y_3) \cos B - (X_1 Y_2 + X_2 Y_1) \cos C = 0$$

Now, the left side of this equation is the polarized form of the expression

$$\Omega = X_1^2 + X_2^2 + X_3^2 - 2X_2 X_3 \cos A - 2X_3 X_1 \cos B - 2X_1 X_2 \cos C$$

Since the polarized form of $2\{yX\}\{zX\}$ is $\{yX\}\{zY\} + \{zX\}\{yY\}$, it follows that any two perpendicular lines through either of two points $(y)$ and $(z)$ are conjugate wo any conic of the form

$$\{yX\}\{zX\} - \lambda \Omega = 0$$

Hence any conic with foci $(y)$ and $(z)$ (see Sec. 9·7) has such an equation; and by varying $\lambda$ we obtain a range of *confocal* conics.

Making $(y)$ and $(z)$ coincide, we obtain the equation

**12·93**                    $$\{zX\}^2 - \lambda \Omega = 0 \qquad\qquad (\lambda > 0)$$

for a circle with center $(z)$. Different values of $\lambda$ yield a self-dual system of concentric circles. The poles of any fixed line wo such a range of circles form a range of points whose parameter $\lambda$ is a linear function of the distance of such a point from the center $(z)$. In fact, as this distance vanishes with $\lambda$, it is actually proportional to $\lambda$. By 9·41, it is also proportional to the square of the radius. But the condition for the circle 12·93 to touch the line $[1, 0, 0]$ is

---

* Salmon (Ref. 37, p. 59).

$$z_1^2 - \lambda = 0$$

Hence, if $z_1$, $z_2$, $z_3$ are the actual distances of $(z)$ from the three sides of the triangle of reference, the circle with center $(z)$ and radius $\rho$ is precisely

$$\{zX\}^2 - \rho^2 \Omega = 0$$

In particular, the incircle (for which $\qquad z_1 = z_2 = z_3 = \rho = r$) is

$$(X_1 + X_2 + X_3)^2 - \Omega = 0$$

or

$$X_2 X_3 \cos^2 \tfrac{1}{2}A + X_3 X_1 \cos^2 \tfrac{1}{2}B + X_1 X_2 \cos^2 \tfrac{1}{2}C = 0$$

or

$$x_1^{\frac{1}{2}} \cos \tfrac{1}{2}A \pm x_2^{\frac{1}{2}} \cos \tfrac{1}{2}B \pm x_3^{\frac{1}{2}} \cos \tfrac{1}{2}C = 0$$

and the circumcircle (for which $\qquad \rho = R \qquad$ and $\qquad z_1 = R \cos A,$ etc.) is

$$(X_1 \cos A + X_2 \cos B + X_3 \cos C)^2 - \Omega = 0$$

or

$$(X_1 \sin A)^{\frac{1}{2}} \pm (X_2 \sin B)^{\frac{1}{2}} \pm (X_3 \sin C)^{\frac{1}{2}} = 0$$

or

$$(aX_1)^{\frac{1}{2}} \pm (bX_2)^{\frac{1}{2}} \pm (cX_3)^{\frac{1}{2}} = 0$$

or

$$a x_2 x_3 + b x_3 x_1 + c x_1 x_2 = 0$$

### EXERCISES

1. Using Cartesian coordinates, identify the absolute involution 12·91 with the degenerate polarity

$$x_1 = X_1, \qquad x_2 = X_2, \qquad x_3 = 0$$

Deduce the condition for lines

$$X_1 x_1 + X_2 x_2 + X_3 = 0 \qquad \text{and} \qquad Y_1 x_1 + Y_2 x_2 + Y_3 = 0$$

to be perpendicular:

$$X_1 Y_1 + X_2 Y_2 = 0$$

2. Obtain the Cartesian envelope equation

$$(z_1 X_1 + z_2 X_2 + X_3)^2 - \rho^2 (X_1^2 + X_2^2) = 0$$

for the circle with center $(z_1, z_2)$ and radius $\rho$.

3. Verify the degeneracy of 12·92.

4. If the trilinear equation $\Sigma\Sigma a_{ij} x_i x_j = 0$ represents a pair of lines, prove that the condition for these lines to be perpendicular is

$$a_{11} + a_{22} + a_{33} - 2a_{23} \cos A - 2a_{31} \cos B - 2a_{12} \cos C = 0$$

5. Prove that the feet of the perpendiculars from any point on the circum-circle to the sides of the triangle lie on a line (Simson line).

6. Find trilinear coordinates for the mid-points of the sides and the feet of the altitudes. Verify that they satisfy the equation

$$(ax_1 + bx_2 + cx_3)(x_1 \cos A + x_2 \cos B + x_3 \cos C)$$
$$- 2(ax_2x_3 + bx_3x_1 + cx_1x_2) = 0$$

or

$$x_1^2 \sin 2A + x_2^2 \sin 2B + x_3^2 \sin 2C$$

$$- (x_1 \sin A + x_2 \sin B + x_3 \sin C)(x_1 \cos A + x_2 \cos B + x_3 \cos C) = 0$$

(the nine-point circle).

7. By examining the involution of conjugate points on the line at infinity $ax_1 + bx_2 + cx_3 = 0$, verify that any circle may be expressed in the form

$$(ax_1 + bx_2 + cx_3)(Y_1x_1 + Y_2x_2 + Y_3x_3) - \lambda(ax_2x_3 + bx_3x_1 + cx_1x_2) = 0$$

8. Show that the incenter, centroid, circumcenter, orthocenter, and nine-point center of the triangle of reference $ABC$ have trilinear coordinates

$$(1, 1, 1), \qquad \left(\frac{1}{a}, \frac{1}{b}, \frac{1}{c}\right), \qquad (\cos A, \cos B, \cos C),$$

$$(\sec A, \sec B, \sec C) \qquad \text{and} \qquad (\cos (B - C), \cos (C - A), \cos (A - B))$$

Verify that the last four all lie on the Euler line

$$[\sin 2A \sin (B - C), \sin 2B \sin (C - A), \sin 2C \sin (A - B)]$$

# APPENDIX I

# THE COMPLEX PROJECTIVE PLANE

Many books have been written on a more elaborate geometry where the coordinates of points and lines, instead of being real numbers, are complex numbers.* A large part of the development of Chap. 12 can be carried over. Even the criterion $\nu/\mu < 0$ for separation (Sec. 12·4) remains significant, provided that we remember that a number cannot be negative without being *real*. (Thus "$\nu/\mu < 0$" no longer means that $\mu$ and $\nu$ have "opposite signs": they might be both imaginary.) The general point collinear with $(a)$ and $(b)$ is $(a + \nu b)$, where $\nu$ is any complex number. Three collinear points

$$(a), \qquad (b), \qquad (a + \mu b)$$

determine a so-called *chain* of points $(a + \nu b)$ where $\nu/\mu$ is real. Thus the line contains infinitely many chains, and instead of 3·31 we have two points decomposing a chain into two segments. Actually *all but one of the axioms of real projective geometry apply also to complex projective geometry*. The single exception is 3·14, which is false if the four points do not all belong to the same chain. In place of that axiom we have the synthetic definition for a chain, as consisting of three collinear points $A$, $B$, $C$ along with the three segments $BC/A$, $CA/B$, $AB/C$. In this terminology, 3·14 states that the chain covers the whole line, which is just what happens in real geometry but not in complex geometry.

Thus everything in Chap. 2 remains valid, but those later results which depend on 3·14 (directly or indirectly) are liable to be contradicted. An interesting example is 3·33, which is contradicted by the following configuration: The nine points defined by the equations

$$x_1^3 + x_2^3 + x_3^3 = x_1 x_2 x_3 = 0$$

* For a good exposition of Hamilton's approach to complex numbers, see Robinson (Ref. 32, pp. 83–84). It is hardly necessary to add that the words *real* and *imaginary* are picturesque relics of an age when the nature of complex numbers was not yet clearly understood.

*viz.*,

$$(0,\ 1,\ -1),\qquad (0,\ 1,\ -\omega),\qquad (0,\ 1,\ -\omega^2)$$
$$(-1,\ 0,\ 1),\qquad (-\omega,\ 0,\ 1),\qquad (-\omega^2,\ 0,\ 1)$$
$$(1,\ -1,\ 0),\qquad (1,\ -\omega,\ 0),\qquad (1,\ -\omega^2,\ 0)$$

where $\omega = e^{2\pi i/3}$, lie by threes on 12 distinct lines in such a way that any two of the points lie on one of the lines. These nine points are the common inflexions of the cubic curves

$$x_1^3 + x_2^3 + x_3^3 + \lambda x_1 x_2 x_3 = 0$$

The classification of projectivities is actually simpler in complex than in real geometry. For now every projectivity has one or two invariant points, and a line cannot fail to meet a conic. On the other hand, besides the projectivity relating $(a + \nu b)$ to $(a' + \nu b')$ there is also an *antiprojectivity* relating $(a + \nu b)$ to $(a' + \bar{\nu}b')$, where $\bar{\nu}$ is the complex conjugate of $\nu$. Both these transformations preserve the harmonic relation. Similarly, there is not only a projective collineation 12·51 but also an antiprojective collineation

$$x_i' = \Sigma c_{ij}\bar{x}_j$$

The projective polarity 12·61 always determines a conic $(xx) = 0$; but there is also an antiprojective polarity

$$X_i = \Sigma a_{ij}\bar{x}_j \qquad\qquad (a_{ji} = \bar{a}_{ij})$$

whose self-conjugate points form an "anticonic" $(x\bar{x}) = 0$.

Many results in real geometry are most easily obtained by regarding the real plane as part of the complex plane, so that a real line appears as a chain on a complex line. This method is especially valuable in the theory of circles; for the absolute involution has two invariant points (the *circular points at infinity*, discovered by Poncelet in 1813), and a circle is most simply defined as a conic that passes through these two points.*

* Salmon (Ref. 38, p. 1).

# APPENDIX II

# BIBLIOGRAPHY

1. APOLLONIUS: *Conicorum* (Heiberg's edition), Leipzig, 1891.
2. BAKER: *Principles of Geometry*, Vol. I, Cambridge, 1929.
2a. ——: *Principles of Geometry*, Vol. II, Cambridge, 1930.
3. ——: *An Introduction to Plane Geometry*, Cambridge, 1943.
4. CHASLES: *Traité des Sections Coniques*, Paris, 1865.
5. COOLIDGE: *A History of the Conic Sections and Quadric Surfaces*, Oxford, 1945.
6. COXETER: *Non-Euclidean Geometry* (2d ed.), Toronto, 1947.
7. ——: *Regular Polytopes*, London, 1948.
8. CREMONA: *Elements of Projective Geometry*, Oxford, 1913.
9. DONNAY: *Spherical Trigonometry after the Cesàro Method*, New York, 1945.
10. DURELL: *Modern Geometry*, London, 1931.
11. ENRIQUES: *Lezioni di geometria proiettiva*, Bologna, 1904. References are made to the French translation, Paris, 1930.
12. FORDER: *The Foundations of Euclidean Geometry*, Cambridge, 1927.
13. ——: *Higher Course Geometry*, Cambridge, 1931.
14. GRAUSTEIN: *Introduction to Higher Geometry*, New York, 1930.
15. HARDY: *A Course of Pure Mathematics* (4th ed.), Cambridge, 1925.
16. HEFFTER and KOEHLER: *Lehrbuch der analytischen Geometrie*, Vol. I, Leipzig, 1905.
17. HESSE: *Vorlesungen über analytische Geometrie des Raumes*, Leipzig, 1897.
18. HESSENBERG: *Grundlagen der Geometrie*, Berlin, 1930.
19. HILBERT: *Grundlagen der Geometrie* (7th ed.), Leipzig, 1930.
20. HODGE and PEDOE: *Methods of Algebraic Geometry*, Cambridge, 1947
21. HOLGATE: *Projective Pure Geometry*, New York, 1930.
22. JOHNSON: *Modern Geometry*, Cambridge, Mass., 1929.
23. LEHMER: *An Elementary Course in Synthetic Projective Geometry*, Boston, 1917.
24. LEVI: *Geometrische Konfigurationen*, Leipzig, 1929.
25. MATHEWS: *Projective Geometry*, London, 1914.
26. O'HARA and WARD: *An Introduction to Projective Geometry*, Oxford, 1937.
27. PASCH and DEHN: *Vorlesungen über neuere Geometrie*, Berlin, 1926.
28. PIERI: I principii della geometria di posizione, composti in sistema logico deduttivo, *Memorie della Reale Accademia delle Scienze di Torino* (2), vol. 48, pp. 1–62, 1899.

29. PONCELET: *Traité des propriétés projectives des figures* (2d ed.), Vol. I, Paris, 1865.

30. ———: *Traité des propriétés projectives des figures* (2d ed.), Vol. II, Paris, 1865.

31. REYE: *Die Geometrie der Lage* (6th ed.), Leipzig, 1923.

32. ROBINSON: *The Foundations of Geometry*, Toronto, 1940.

33. ROBSON: *An Introduction to Analytical Geometry*, Vol. I, Cambridge, 1940.

34. ———: *An Introduction to Analytical Geometry*, Vol. II, Cambridge, 1947.

35. RUSSELL: *Introduction to Mathematical Philosophy* (2d ed.), London, 1930.

36. ———: *The Principles of Mathematics* (2d ed.), London, 1937.

37. SALMON: *A Treatise on Conic Sections* (6th ed.), London, 1879.

38. ———: *A Treatise on the Higher Plane Curves* (3d ed.), Dublin, 1879.

39. SMITH: *An Elementary Treatise on Conic Sections*, London, 1921.

40. VON STAUDT: *Geometrie der Lage*, Nuremberg, 1847.

41. ———: *Beiträge zur Geometrie der Lage*, Nuremberg, 1857.

42. VEBLEN and YOUNG: *Projective Geometry*, Vol. I, Boston, 1910.

43. ——— and ———: *Projective Geometry*, Vol. II, Boston, 1918.

44. WHITEHEAD: *The Axioms of Projective Geometry*, Cambridge, 1906.

45. YOUNG: *Projective Geometry* (Fourth Carus Monograph), Chicago, 1930.

# INDEX

(References to pages; principal references in boldface)